The Interstitial Spaces of Urban Sprawl

This book proposes the idea of interstitial space as a theoretical framework to describe and understand the implications of in-between lands in urban studies and their profound transformative effects in cities and their urban character.

The analysis of the interstitial spaces is structured into four themes: the conceptual grounds of interstitial spaces; the nature of interstices; the geographical scale of interstices; and the relationality of interstices. The empirical section of the book introduces seven cases that illustrate the varied nature of interstitiality to finally discuss its implications in the broader field of urban studies. Reflections upon further lines of enquiry and theories of urbanisation, urban sprawl, and cities are highlighted in the conclusion chapter.

This is the ideal text for scholars of urban planning, strategic spatial planning, landscape planning, urban design, architecture, and other cognate disciplines as well as advanced students in these fields.

Cristian Silva is an architect and urbanist with an MA in Architecture from the Pontificia Universidad Católica de Chile and a PhD in Urban Studies from the Bartlett School of Planning, University College London (UCL). He is currently teaching in the areas of institutional and policy context of planning practice, urban design, health and wellbeing, and independent research at Queen's University Belfast, Northern Ireland, UK. His research profile lies in the intersections between urban design, spatial planning, and social theory.

Routledge Contemporary Perspectives on Urban Growth, Innovation and Change

Series edited by **Sharmistha Bagchi-Sen**, Professor, Department of Geography and Department of Global Gender and Sexuality Studies, State University of New York-Buffalo, Buffalo, NY, USA and **Waldemar Cudny**, Associate Professor. Working at The Jan Kochanowski University (JKU) in Kielce, Poland.

Urban transformation affects various aspects of the physical, social, and economic spaces. This series contains monographs and edited collections that provide theoretically informed and interdisciplinary insights on the factors, patterns, processes and outcomes that facilitate or hinder urban development and transformation. Books within the series offer international and comparative perspectives from cities around the world, exploring how 'new life' may be brought to cities, and what the cities of future may look like.

Topics within the series may include: urban immigration and management, gender, sustainability and eco-cities, smart cities, technological developments and the impact on industry and on urban societies, cultural production and consumption in cities (including tourism, events and festivals), the marketing and branding of cities, and the role of various actors and policy makers in the planning and management of changing urban spaces.

If you are interested in submitting a proposal to the series please contact Faye Leerink, Commissioning Editor, faye.leerink@tandf.co.uk.

Place Event Marketing in the Asia Pacific Region
Branding and Promotion in Cities
Edited by Waldemar Cudny

Post-socialist Shrinking Cities
Edited by Chung-Tong Wu, Maria Gunko, Tadeusz Stryjakiewicz and Kai Zhou

Growth and Change in Post-socialist Cities of Central Europe
Edited by Waldemar Cudny and Josef Kunc

The Interstitial Spaces of Urban Sprawl
Geographies of Santiago de Chile's *Zwischenstadt*
Cristian A. Silva

The Interstitial Spaces of Urban Sprawl

Geographies of Santiago de Chile's *Zwischenstadt*

Cristian Silva

LONDON AND NEW YORK

First published 2022
by Routledge
2 Park Square, Milton Park, Abingdon, Oxon OX14 4RN

and by Routledge
605 Third Avenue, New York, NY 10158

Routledge is an imprint of the Taylor & Francis Group, an informa business

British Library Cataloguing-in-Publication Data
A catalogue record for this book is available from the British Library

Library of Congress Cataloging-in-Publication Data
A catalog record has been requested for this book

ISBN: 978-0-367-33471-0 (hbk)
ISBN: 978-1-032-17071-8 (pbk)
ISBN: 978-0-429-32001-9 (ebk)

DOI: 10.4324/9780429320019

Typeset in Times NR MT Pro
by KnowledgeWorks Global Ltd.

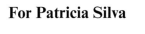

For Patricia Silva

Contents

List of Figures

Foreword

With the possible exception of Andrea Brighenti's (2016) edited volume *Urban Interstices*, the academic and popular conversation regarding what Cristian Silva here consolidates under the term 'interstitial spaces' is strewn across a number of disciplines (sociology, architecture, geography, urban design, urban planning), and under a number of competing terms (*terrains vague*, interfragmentary spaces, vacant lands, open spaces, undeveloped areas, and others) that have tended to ensure that this phenomenon has not received the attention it deserves.

Our built environments are highly porous – more than we have cared to acknowledge in professional and academic terms, even if the opportunities presented by such porosity have long been seized by ordinary citizens across the global north and marginalised populations across the global south. The interstitial spaces that lie within and between metropolitan areas exist in much more variety than commonly presented in the most conspicuous instances of urban and industrial decay and abandonment. These are merely some of the largest interstitial spaces in geographic scale but – as Cristian Silva makes apparent here and in his previously published articles – they exist at a variety of scales from the macro-, or intermetropolitan, to the micro- or architectonic.

Nor, on closer inspection, are these interstices abandoned entirely. They are often in some use – even if that use is not what would commonly be understood as highest and best in real estate investment and management terms. Moreover, they are spaces upon which imaginations continue to be operative. Far from non-places – as places of things remembered they are for some sections of the population, thick with a sense of community or identity – the interstitial spaces are foci for imaginaries of future possibilities. Whether this be as sources of profit from the most lucrative forms of real estate development or the sources of wellbeing and social value often sought by local citizens, the interstitial spaces can be analysed, that is, in terms of their uses and their desired uses.

Cristian Silva brings the eye of a trained and by now well-travelled architect to the discussion of interstitial spaces. To add to this, his doctoral research – engaging more fully with planning, urban and sociological theory – is a visually and graphically sensible piece that wedded very fruitfully with more abstract theoretical concerns on interstitial spaces and the extensive nature of contemporary urbanisation. The book is triply integrative: of Cristian's own doctoral and published shorter works; of perspectives drawn from across a variety of disciplines including planning, architecture, urban and landscape design, and geography and development studies; and in the understanding of urban interstices as both bounded spaces but also inherently relational while engaging the attention of different societal interests and their potential connective role within and between cities. The result is a uniquely original analysis and one that deserves the full attention of urban scholars.

Cristian exemplifies his ideas concerning interstitial spaces with reference primarily to Santiago de Chile, drawing attention to the fascinating capital city less studied in the vast volumes of urban literature. As well as being home to the typical infrastructure that has generated interstitial spaces found in every city, Santiago de Chile is also home to spectacular stand-offs – as in the old Cerrillos airport – between interested parties over the re-use of 'vacant' sites that compare in their scale and duration to those found anywhere in the world. Santiago de Chile is also full of curious mixes of political ideologies and associated planning policies that have conspired to produce a variety of urban interstices. However, the discussion also draws on examples from Europe to make clear the global relevance of interstitial spaces across a variety of societal and economic system contexts.

I have had the pleasure of working with Cristian as he has developed his unique and original perspective on cities. I am glad that you now have the opportunity to enjoy reading this product of a lively and inquisitive mind.

Nicholas A. Phelps
Melbourne

Preface

The idea of 'interstitial space' has been at the centre of my scholarly work since finishing my studies of architecture in the 2000s. Once I got my first position in academia, I started a few projects on housing and urban informality in which the focus of my research was precisely on the space of relation between the informal and the formal city: the interstitial space. In the south of Chile these relations are peculiar, as what we tend to identify as an 'informal' settlement – under the formal planning nomenclatures – is many times a fishing or rural village that has been engulfed by the sprawling growth of the institutionally planned city. These villages are not 'slums' as such, but informally produced neighbourhoods organised by the mechanism of survival different from those based on logics of land privatisation, regulated buildings, and bi-dimensional zoning. The encounters between the formal and informal cities are sometimes marked by tense interactions with central authorities who aim to eradicate the villages. The inhabitants of these fishing 'slums', however, tend to resist politics of eradications, regulations, or the construction of standardised homes arranged by planning authorities. In the end, the interstitial space is always the only cushioning zone – a sort of invisible wall – that enables the interaction between dissimilar realities, liminal boundaries where differing realities and mechanisms of the production of the space coexist.

While pursuing my Master's degree in Santiago de Chile, I continued studying the spaces of interaction between the formal and informal city. In this experience, however, the slums were produced by large rural–urban migrations and the illegal occupation of private lands within the city. Here, the interactions between the families of the slums and institutional actors were politically and economically mediated, as the inner lands occupied by the slums were well located for real estate projects. So, the informal occupants were eradicated to the peripheries, and the lands enabled for urban projects. The interstitial spaces of encounter become abandoned, finally encroached by others who used them as landfills or simply further urbanised. I observed that the interstitial spaces can be useless as spaces of mediation and can even disappear.

I decided then to go deeper in understanding the varied characteristics, temporalities, magnitudes, and relationalities of interstitial spaces beyond formal-informal dichotomies, and thus, during my PhD at UCL (UK), I developed this agenda of research that would hopefully serve as basis for further studies on urban sprawl and interstitial spaces. While doing the PhD, I also visited more than forty cities of very different countries – including capital cities and regional towns in Europe, the United States, Russia, and Latin America. In my journeys I found a considerable range of interstitial spaces of different sorts that constitute a significant proportion of the geography and history of cities. In Athens and Rome, for instance, some interstitial spaces speak about thousands of years of western history; in Saint Petersburg and Moscow they speak about the production of colossal magnitudes of energy; in Berlin the interstitial spaces still show some scars of the wars, while in Manchester they highlight the heydays of the industrial revolution. In Helsinki and Stockholm, huge rocks and sea channels speak about the millennial geology of our planet while separating suburban neighbourhoods where we walk every day. In New York, interstitial spaces show the potential of our imagination and how quickly we can transform the space, while in Latin America they become shared spaces occupied in very creative ways. During my stay in New Zealand, I gathered further insights on interstitial spaces that have been then collated with other experiences in Belfast, Northern Ireland, and I can confirm that interstitial spaces have salience to the vast majority of cities. As such, they configure a fascinating geography that shows alternative views of what makes cities urban, a geography that is nevertheless shadowed by more contingent urban problems even when some of the most pressing issues of our societies occur in the system of interstitial spaces.

The interstitial spaces are not a simple collection of geographies but rather a scope of knowledge that deserves more attention if we want to understand cities and the urbanisation process in a more comprehensive way. This book is an attempt to contribute to this.

Cristian A. Silva
Belfast

Acknowledgements

I would like to thank Prof Nicholas Phelps, who supervised the PhD research that served as the main basis for this work. His contribution to the literature of cities and processes of urbanisation has been inspirational.

I would also like to thank the School of Natural and Built Environment, Queen's University Belfast, for supporting this work through the Grant D820PAC, and the National Commission of Science and Technology of Chile (CONICYT), Grant 72110038.

Finally, I wish to thank my family, particularly my wife, Marcela, for her loving support.

Interview sources

1 Director of the School of Architecture, University of Cambridge, 16 April 2014
2 Architect in charge of the Social Housing Plan for the Presidential Programme of Michele Bachelet (2014–2018), 14 March 2014
3 National Director of Urban Development, Ministry of Housing and Urbanisation (MINVU), Chile, 05 May 2014
4 Director of the Department of Regional Planning, Metropolitan Regional Government of Santiago (GORE), 06 May 2014
5 Urban Designer in charge of 'Ciudad Parque Bicentenario' (CPB), SERVIU, MINVU, 07 May 2014
6 Architect in charge of 'Ciudad Parque Bicentenario' (CPB), SERVIU, MINVU, 07 May 2014
7 Researcher, Institute of Territorial and Urban Studies (IEUT), Universidad Católica de Chile; former Director of 'Ciudad Parque Bicentenario' (CPB), MINVU, 08 May 2014
8 Professional at the National Service of Environmental Evaluation. Ministry of Environment, 09 May 2014
9 Urban Designer in charge of 'Parque Bicentenario', CPB Project, Montealegre-Beach Architects, 09 May 2014
10 Director and member of URBE Consultants, 12 May 2014
11 Director of research and development of La Platina Research Centre, National Institute of Agricultural Research (INIA), Ministry of Agriculture, La Pintana, 12 May 2104
12 Advisor at National SEREMI of Agriculture (2010–2014), 13 May 2014
13 National Secretary of Agriculture (SEREMI) (2010–2014), 13 May 2014
14 Professional at National Service of Environmental Evaluation. Ministry of Environment, 14 May 2014
15 National director of planning at Ministry of Public Works (MOP). Former Director of 'Ciudad Parque Bicentenario', MINVU, 14 May 2014
16 Director of 'Ciudad Parque Bicentenario – CPB' Project, SERVIU, MINVU, 14 May 2014

17 Former Director of 'Ciudad Parque Bicentenario' Project at MINVU (2001–2004), 15 May 2014

18 Architect in charge of Infrastructural Development, Campus Antumapu, Faculty of Veterinarian Sciences, Universidad de Chile, 16 May 2014

19 Coordinator of Housing and Real Estate Development at Chilean Chamber of Construction (C.Ch.C.), 19 May 2014

20 Deputy of district N°20, communes of Cerrillos, Estación Central and Maipú. 22 May 2014

21 Senator for the VIII Circunscripción de Santiago Oriente, 22 May 2014

22 General Manager of Urban Studies at Chilean Chamber of Construction (C.Ch.C.), 22 May 2014

23 Coordinator of Territorial Management, Department of Urban Studies, Chilean Chamber of Construction (C.Ch.C.), 22 May 2014

24 Director of urban planning, Municipality of La Pintana, 23 May 2014

25 Director of urban planning, Municipality of Cerrillos, 26 May 2014

26 Director of urban planning, San Bernardo Municipality, 26 May 2014

27 Honorary advisor and real estate developer, Chilean Chamber of Construction (C.Ch.C), 27 May 2014

28 Director of urban planning, Municipality of Lo Espejo, 28 May 2014

29 Director of urban planning, Municipality of El Bosque Municipality, 28 May 2014

30 Director of urban planning, Municipality of La Florida, 05 May 2014

31 Architect and former General Manager of Aeronautic Studies, Cerrillos Airport, Ministry of Public Works (MOP), 29 May 2014

32 Director of urban planning, Puente Alto Municipality, 30 May 2014

33 Former Minister of Housing, Urbanization and Public Lands (2001–2004), 03 June 2014

34 Director of urban planning, Municipality of Pedro Aguirre Cerda, 04 June 2014

35 Director of urban planning, Municipality of Padre Hurtado, 04 June 2014

36 Director of Community Organizations at Municipality of La Pintana, 10 June 2014

37 Director of the Department of Environmental Operations. Municipality of La Pintana, 10 June 2014

38 Funder Member of the Union 'Huertos Obreros y Familiares' [Worker and Familial Orchards]. Resident of 'Huertos Obreros', La Pintana, 10 June 2014

39 President of the Union 'Huertos Obreros y Familiares' [Worker and Familial Orchards]. Resident of 'Huertos Obreros', La Pintana, 10 June 2014

40 President of the Water Union 'Villa Las Rosas', La Pintana, 10 June 2014

41 Director of irrigation and member of the Agricultural Cooperative 'José Maza', 'Huertos Obreros y Familiares' [Worker and Familial Orchards], La Pintana, 10 June 2014

42 President and member of the Agricultural Cooperative 'José Maza', 'Huertos Obreros y Familiares' [Worker and Familial Orchards], La Pintana, 10 June 2014

43 Member of the Agricultural Cooperative 'José Maza', 'Huertos Obreros y Familiares' [Worker and Familial Orchards], La Pintana, 10 June 2014

44 Researcher at Department of Urban Planning, Universidad de Chile, 12 June 2014

45 Consultant in local food systems at Food and Agricultural Organisation (FAO-RLC), UN, Chile, 12 May 2014

46 Director of the NGO 'Planta-Banda', Espacio Padre Mariano. Municipality of Providencia, 12 May 2014

47 President of the committee of neighbours 'Villa San Ambrosio III', La Pintana, 13 June 2014

48 Secretary of the committee of neighbours 'Villa San Ambrosio III', La Pintana, 13 June 2014

49 Resident of Villa San Ambrosio III. Police Officer at La Pintana, 13 June 2014

50 Resident of Villa Ambrosio III, La Pintana. 13 June 2014

51 Resident of Villa San Gabriel, La Pintana. 13 June 2014

52 Resident of Villa San Gabriel, La Pintana, 13 June 2014

53 Intendant of the Metropolitan Region of Santiago, 20 June 2014

54 Director of the School of Construction, Universidad de la Américas. 23 June 2014

55 Director of Environmental Management, Municipality of La Pintana, 25 June 2014

56 Director of urban planning, Municipality of Maipú, 22 May 2014

1 Enquiring into urban sprawl

Is it all about the built-up space?

1.1 Introduction

Against all predictions, the expansion of processes of urbanisation and the sprawling development of cities and regions have become a global phenomenon that remains as one of the most longstanding patterns of urban development. It is not possible to understand cities and their development without paying particular attention to urban sprawl – a permanent structure of cities since the years after WWII – and the type of urban development it suggests. The magnitude of sprawling urban development has been partly fuelled by increasing migrations into cities at a scale and speed with no precedents in human history. Receiving about 'three million people per week' worldwide, cities are impacted by sharp pressures on food production, energy supply, infrastructure, housing, services, and employment growth (UN, 2018); 'a recognition that the exceptional rate of city expansion into larger (mega) city-regions continues apace' (Harrison & Heley, 2015, p. 1114). This has profound transformative effects on the spatial character of cities and urbanised regions while somehow it confirms the almost unquestionable perception that 'we live in an urban era in the sense that cities formally represent the principal geographic containers within which contemporary human society unfolds' (Storper & Scott, 2016, p. 117).

This urbanisation process, however, has been questioned by the undeniable fact that what seems to be an urban world is really the manifestation of an extended, dispersed, fragmented, diverse, and multifaceted process of suburbanisation in which urban sprawl goes rampant all across developing and developed countries alike (Abubakar & Doan, 2017; Bruegmann, 2005; Kennedy et al., 2016). While the massive wave of present urbanisation is often referred to as an 'urban revolution', most of this startling urban growth worldwide is happening at the margins of cities. Apposed to most orthodoxies, two-thirds of the world's population really live in 'suburban' areas, and more than 80% of urban dwellers in many cities are already 'suburban' (Keil, 2017). It is not a surprise that Roger Keil wonders *what is not suburbia* while questioning the usual reading of the urban in most theories of the city. The magnitude of suburban sprawl is changing the nature of cities and

DOI: 10.4324/9780429320019-1

challenging our cognition about their essential components, structures, and interrelations. From monocentric cities characterised by binomial centre-periphery relations, we have seen the emergence of more polycentric, multifunctional, dispersed, and post-suburban regions characterised by the diversity of suburbia and a renewed linkage between workplaces and residences (Burger & Meijers, 2012; Fishman, 1987; Phelps & Wood, 2011). It is clear that Cedric Price's scramble egg metaphor of modern cities would find today more than a single case to be matched with. The meanings and the way of how city-centres, peripheries, the urban-rural transect, suburbia, peri-urban, and fringe/belt areas are manifested are not possible to be apprehended by existing analytical conventions (Phelps, 2015). So, urban sprawl has become a challenging geography that remains as one of the most intractable problems of our modern society; a geography that calls for the development of urban theories and experiments in planning beyond normative rationales and bi-dimensional conceptions of land use (Gallent & Shaw, 2007); a phenomenon that we still should look at, and should learn from.

Having been extensively examined in terms of origins, evolution, meanings, impacts, and its underlying politics (Barnes et al., 2001; Barrington & Millard, 2015; Chen et al., 2018; Phelps, 2012), urban sprawl has been the subject of a strong 'city-oriented' focus ant thus, examined through the lenses of an observable and analytically plausible dimension: *the built-up space*. There is no surprise that the built-up dimension of urban sprawl has been the ultimate object of study as it directly contributes to our understanding of cities and our positivist conceptions of the 'production of the space' (Brenner & Elden, 2009). However, the sprawling character of cities has been 'exploding relentlessly beyond their boundaries, producing a highly uneven urban fabric that ceaselessly extends its borders across non-urban geographies' (Arboleda, 2016, p. 234). The interchange of goods across transitional spaces of connectivity – such as the conurbation zone between cities, towns, and infrastructures – are all part of a regional in-between space of networks that support the economy of cities and contribute to their urban life (Kassa, 2013; Mazzoni & Grigorovschi, 2015). So, extensive urban sprawl illustrates a magnitude of urbanisation that does not always take place in cities but beyond; in-between spaces that are not celebrated as wonders of cities and remain invisible from the view of urban theories. It is clear that urban centres have been the core of economic growth and seminal political processes, but to gain a better understanding of the urban condition is essential to look at the non-urban geographies that lie between urban centres, enclaves, and other forms of urbanisation based on the 'built-up' as the ultimate expression of societal development; there is a clear lack of attention to the transformative processes that occur in the interstitial spaces between the built-up space of cities, and between metropolitan areas that are part of the sprawling expansion of cities and regions shaped by complex transcalar processes.

The significance of these non-urban geographies not only relates to their spatial scale but also to their functional properties. Some of these interstitial spaces overpass the regional hinterland while supporting extensive chains of transcalar production (and reproduction) of the built environment. Identified as 'colossal landscapes' (Arboleda, 2016, p. 234) – considering their economic, environmental, and global impacts – some of these spaces play a critical role in the accumulation of capital at the point of becoming 'planetary' hubs of economic influence (Brenner & Schmid, 2014; Scott & Storper, 2014; Shaw, 2015). This is the case of large mining sites that supply international markets and contribute to the sprawling growth of far afield cities and regions; a process that vanishes the boundaries of where 'the urban' ends and connects with other regions that make the interchange a transnational exercise (Da Costa Braga & Rigatti, 2009; Zook & Sheltom, 2013). In a closer scale, the informal commerce deployed in the interstices of many suburban areas speaks about the liveability of cities aside from the formally institutionalised production and design of the space. Other interstices are informally occupied by marginal groups that use them as trenches against institutional forces of normalisation (Shaw & Hudson, 2009). These interstices evince the spontaneity of human processes that clashes with planned functions; a debate highlighted by Tschumi's ideas of 'disjunction' (Tschumi, 1996). From the 'colossal landscapes' to the liminal interstices that flourish in the inner lattice of the urban, these spaces highlight the scales at which they can manifest and the piecemeal opportunities distributed within urban politics of control (Phelps, 2012).

Although it has been affirmed that we live in an urban era in the sense that most people live in cities (with undeniable statistics about), the question of what lies in the non-urban geographies has gained traction over the last years as not all aspects of life are (and have been) developed in cities. It has been clarified that a considerable portion of economic growth rests on the geographies found beyond the space occupied by cities, and in the spaces between built-up areas (Harrison & Heley, 2015). This does not mean, however, that processes of economic development that take place outside (or between) cities – or in their inner interstices – occur by default in the rural countryside or in mere leftover spaces. The notion of a 'rural' realm would be somehow reductive as these large non-urban geographies are as diverse as urban sprawl itself, and indeed, define its substantial character as a fragmented territory. Pieces of countryside, deserts, forests, mountains, ranges, farming lands, industrial facilities, landfills, mining sites, undeveloped areas and open tracts of different sorts, infrastructural spaces, protected ecological reservoirs, conurbation zones, regional parks, leftover spaces, derelict lands, boundary areas, and others, configure an eclectic interstitial geography where different societal processes unfold. Some of these spaces still remain 'wild' or untouched, but many others describe different levels of planning and political mediations; mediations that highlight the character of the urban as not only confined to 'cities'. Somehow, the supposedly pristine or natural world between cities and their built-up spaces is indeed 'an extent to which'; a realm where different infrastructures of linkage and modes of urbanisation speak about the commodification of natural resources, land, and the integration of goods and services into the voracious environment of cities. All these processes have undoubtedly opened critical enquiries in urban anthropology, urban planning, geography, urban political ecology, urban studies, and other disciplines engaged with prospecting the built environment, and have expanded their imaginaries of 'the urban' beyond urban agglomerations (Angelo & Wachsmuth, 2015; Heynen et al., 2006). In that sense, Lefebvre's (1970) proposition that we are evolving towards an 'urban society' would only make sense if these non-urban geographies are included as part of, and by questioning the underlying premise that the society and the city are becoming the same entity.

The fact that the wide spectrum of interstitial spaces that compose the non-urban geographies within, between and beyond the sprawling expansion of cities challenges the usual readings of the urban in most theories about the city is one of the main starting points for this book. These in-between spaces in which the urban unfolds are named *'interstitial* spaces'; a dimension of sprawling growth that questions the conceptual apparatus from where we interpret cities and their substantial components. The fundamental point is that cities – and specifically sprawling growth – cannot be understood without reference to the interstitial spaces that are produced alongside the urbanisation process and in which a substantial content of 'the urban' is produced. As such (and although aside from cities or neglected by), these interstitial spaces need to be acknowledged as an integral component of the urban condition.

The term *interstitial space* has been chosen because it conceptually encapsulates the in-between nature of the urban condition while providing a spatial and territorial sense of 'where' the urban unfolds. An interstitial space (although initially abstract) suggests a concrete form of activity (even when apparently empty) and is somehow modelled by its surroundings in spatial and functional terms. Although these interstitial spaces are sometimes referred to as 'in-between' spaces, the term 'in-between' remains spatially insipid (what is the form of an 'in-between' entity, what is this for?), and structurally delinked from its surroundings. An 'in-between' can be a third (independent) entity between two objects that creates further in-between spaces, but an *interstitial space* is a space of relations; a space that takes its meaning in regard of 'where' – or in-between what – it is placed. So, an *interstice* is a context-dependent space that is never delinked from its surroundings, although these can be further away from immediate spatial boundaries. A tree between two houses, for instance, is an in-between entity, but the space where the tree is located is the *interstitial space*. As such, the interstitial space is modelled by the tree and its surrounding houses that define the interstitial space as a context in itself.

At an urban scale, this is why an urban forest is more an interstitial space rather than a mere 'in-between' entity, as an urban forest not only talks about its own ecological characteristics but also the surroundings that shape the conditions in which the forest grows and the non-immediate surroundings that shape its wider ecological functions. Thus, the scale of an interstitial space can be far beyond the space occupied by the physical elements that compose its interstitial condition. A mining site that occupies a certain portion of territory in the countryside – although physically isolated from the cities that surround it – is nevertheless linked to wider ecologies and the economy of cities through different infrastructures and global networks of international trade. These are 'colossal landscapes' (Ernstson, 2021; Arboleda, 2016) that talk about the extended scales at which the interstitial spaces can manifest. In that sense, an *interstitial space* – by contrasting or assimilating information from its surroundings – shapes its own identity as part of the urban. These conceptual disquisitions will be developed in detail in the theoretical Chapter 3 of this book.

Although identified, this interstitial geography is often overlooked or defined as empty, undeveloped, residual, marginal, under-developed, isolated, randomly produced, vacant, or simply inert. Nothing could be further from the truth, however, as 'when one penetrates the system of interstitial spaces and starts to explore it, one realises that what has been called "empty" is not so empty after all. Instead, it contains a wide range of uses' (Sousa Matos, 2009, p. 66). The interstitial spaces are realms of unexplored integration, socialisation, and encounter, and configure a parallel urban system that operates in the shadows of our mainstream narratives of the 'urban'. The interstices also confirm the transitional nature of cities while being 'pending lands' in the process of eventually becoming built-up. This suggests alternative temporalities compared to their urbanised counterpart – the built-up space – and highlights the inseparability of analytical perspectives and normative agendas on urban dynamics as much as the built environment. The interstitial geography is a spatial system that can also be distinguished in terms of its own specific dynamics and characteristics, a spatial system in its own right.

In this book, the notion of *interstitial space* is developed as a theoretical framework to describe, categorise, understand, and operationalise the analysis of the non-urban geographies of cities and regions as outcomes of sprawling development and their implications in urban studies. The sprawling development of cities and regions clarifies the expanding connotation of the term 'urban sprawl' at local, regional, and global scales, and suggests reflections beyond its negative connotations as the context of the interstitial spaces. As such, interstitial spaces offer an alternative point of entry into the studies of urban sprawl and the urbanisation process and force inter-disciplinarity in the understanding of city regions and the elements that make them 'urban'.

1.2 Research design

The material presented in this book is partly distilled from a PhD research conducted at UCL (UK) between 2012 and 2017 on interstitial spaces of urban sprawl that used as a case study in the capital city of Chile, Santiago. Findings of subsequent research of urban sprawl in the Australasian context, Europe and Latin America have been included. Fractions of this material have been published in various journals and presented in conferences in the United States, Europe, Australia, and New Zealand, which are also part of this book. In particular, there is a paper published in 2017 in the *Urban Studies* Journal – co-authored with Prof Nicholas Phelps – that reflects the core of this PhD research in which the research agenda for the study of interstitial spaces is presented in a synthetic way. The last collection of empirical data has been supported by the School of Natural and Built Environment at Queen's University Belfast. As part of the theoretical discussion around suburban interstices, a few examples are provided from different places – including Belfast – although most of the evidence presented in the empirical section of this book comes from the capital city of Chile – Santiago.

The argument also steps on previous works on urban sprawl and interstitial spaces. The discussion around urban sprawl includes closely related literature on suburbanisation, post-suburbanisation, peri-urbanisation and transformation of fringe belt areas. This allows understanding the magnitude of urban sprawl and its salience at different geographical scales to include the interstitial spaces as part of. These bodies of research are combined with those around the interstitial spaces – that appear as quite fragmented and singular in the treatment of interstitial spaces in architectural, ecological, or other terms – and the theoretical gateways from spatial planning, urban political ecology, anthropology, political theory, and geography. This allows the understanding of interstitial spaces from an interdisciplinary perspective that places them as part of the urbanisation process in spatial, social, economic, political, and environmental terms (Figure 1.1).

Santiago de Chile represents most Latin American cities and is a good example of the significance of interstitial spaces in contexts of urban sprawl specifically. Santiago is a capital city that has been the subject of aggressive housing policies in the last 40 years with important consequences for social and spatial segregation (Borsdorf et al., 2007). The fragmented suburban expansion of Santiago is linked to different modes for the production of space, including infrastructural lands, subsidised residential zones, and privatised construction of housing and infrastructure. The city also shows informally occupied areas of 'autoconstruction', unregulated conurbations and a dispersed expansion that has spanned outer districts and satellite towns as part of metropolitan development. For some, this fragmentation illustrates planning rationales adjusted to facilitate urban growth, land privatisation, and centralisation of social housing supply, all elements of a long-standing neoliberal planning agenda (Vergara & Boano, 2020). This

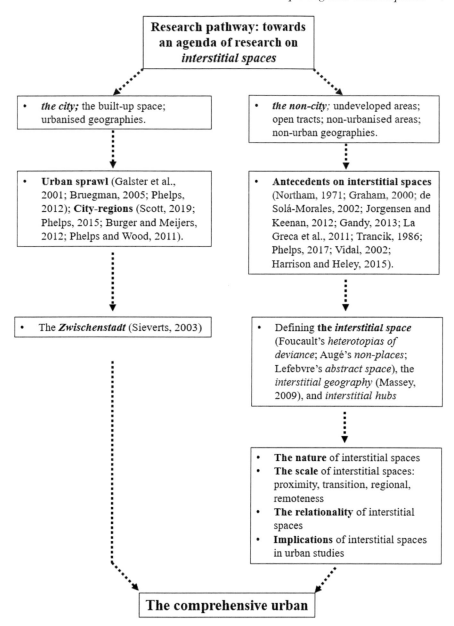

Figure 1.1 The theoretical entries of interstitial spaces of urban sprawl.

case is used as a basis for abstraction and generalisation (Monkkonen et al., 2018) as the interstitial spaces have salience to the vast majority of cities. In this context, seven interstitial spaces are examined; cases identified as strategic by different actors that represent the varied array of functional categories of interstitiality including infrastructural lands, abandoned sites, farming areas, industrial facilities, lands of financial speculation, and inorganic conurbation zones inter alia. The case of Santiago de Chile and its interstitial spaces is expanded in Chapter 7.

The methodological approach of this study includes a multidisciplinary literature review and assessment of the diverse literature on urban sprawl and interstitial spaces. Many academic fields are relevant for such a multidisciplinary review, including economic geography, architecture, urban design, environmental sciences, landscape planning, landscape ecology, spatial planning, policy analysis, political science, anthropology, sociology, and urban studies. This also includes the systematic review of recent books, journals, and government reports, conference proceedings, websites, and online publications that focus on issues related to growth management and interstitial spaces. As a study-case research (Yin, 2009), a mixed methodological approach was used to articulate qualitative and quantitative data (Tashakkori & Creswell, 2007). The analytical approach is based on the examination of policy conflicts and stakeholders' understandings of Santiago's urban development, which allowed figuring out how planning policies give effect to suburban sprawl in the case of Santiago and the appearance of interstitial spaces. The selected interstitial spaces ensure conclusions on the respective strengths of using a multi-study case approach in contributing to knowledge (Burawoy, 1991; Flyvbjerg, 2006).

Document review (Bowen, 2009) considered secondary research and institutional policies, legal norms, central government plans and regulations, urban design projects, local and central development plans, and historical records covering Santiago's urban growth. Thematic analysis and critical review of these documents allowed identifying key themes around the emergence of interstitial spaces in Santiago (Gavin, 2008). This information was collated with empirical data from official statistical databases and 56 semi-structured interviews – conducted between 2014 and 2016 – with a range of actors that included planners and designers of large urban projects in Santiago, policy-makers, and politicians (central/local), metropolitan authorities, social and environmental organisations, residents, academics and private developers, all selected for their first-hand knowledge of Santiago's urban development (Galletta, 2013; Stender, 2017). Interviews were designed in accordance with the sponsors' ethical codes of conduct; key informants are anonymised, so that the respondents could be frank, without fear of social, professional, or political repercussions. Thematic analysis was implied to codify data and frame the analytical structure and the research aims (Evans & Lewis, 2018).

Site visits and direct observations were conducted between 2014 and 2016 to assess the spatial quality and ongoing suburbanisations and interstitial spaces in Santiago. Considering the selection of multiple case studies, observations were based on Rayback's approach, which suggests visual records (photographs) and measurements to establish commonalities and differences and corroborate if an area of analysis meets the study goals (Rayback, 2016). Photographs provided data on the morphological and material composition of Santiago's suburban sprawl and its interstices. These are used as evidence – specifically, on physical infrastructure, environmental and spatial attributes, land uses, day-to-day activities, accessibility, and quality of surroundings – and to contrast maps and written records (Collier & Collier, 1986; Roberts, 2016). The morphological analysis included mapping the suburban expansion of Santiago and interstitial spaces over the last 40 years following a Conzenian approach where building patterns, open space, land subdivisions, streets, vegetation, and land land-uses were documented (Kropf, 2011).

1.3 Statement of aims

This book discusses the idea of sprawling urbanisation from its interstitial dimension. The book's main argument is that the sprawling development of cities and regions is composed of not only built-up lands but also *interstitial spaces* that lie between developments. These interstitial spaces emerge as critical components of the urban condition, although they have received little attention in the extant theories about the urban. This lack of attention is partly explained by the almost exclusive focus on urban manifestations as a synonym of the *built-up*. The interstitial spaces that compose the geography between built-up areas, cities, and regions range from liminal spaces found within the sprawling urban expansion up to the large landscapes between (and beyond) metropolitan areas. These interstitial spaces are the context of significant transformations, especially as they are the scenarios of radical economic changes and environmental alterations linked to the rhythms and cultures of the modern metropolis and extended processes of planetary urbanisation. On this basis, it is argued that interstitial spaces have profound transformative effects in cities and their urban character despite the fact that the analysis of such interstices has been quite fragmented and marginal in the elaboration of theories around the city and the urban.

The analysis of the interstitial spaces is structured in four themes:

a *The conceptual grounds of interstitial spaces:* Interstitial spaces are heterotopic and abstract non-places. As such, they encapsulate and unify a fragmented literature around the spaces 'in-between'. The interstitial spaces compose the *interstitial geography* and support the emergence of *interstitial hubs*.
b *The nature of interstices:* Interstitial spaces emerge as planned spaces and also as random outcomes of less controlled processes in planning. As such, they describe different temporalities of preservation and change while having economic, social, spatial, and political significance.

   c *The geographical scale of interstices:* Interstices are apparent at multiple geographical scales. These scales comprehend the *interstices of proximity*, *interstices of transition*, *regional interstice*, and *interstices of remoteness*. As such, these scales embrace the wide spectrum of interstitial spaces, including the liminal spaces within the sprawling expansion of cities, up to colossal landscapes of global salience between heavily urbanised regions.
   d *The relationality of interstices*: Interstitial spaces can segregate the city and also operate as connectors between different surroundings. As such, interstitial spaces open further reflections around their infrastructures and network spaces, functions, and spatial components, including border, boundaries, and boundary objects.

The significance of the research relates to the influence of interstitial spaces in urban studies and theories of urbanisation. This is an attempt to shift the attention from the study of self-contained urban enclaves – the city-oriented focus – to the urban spaces that lie in-between built-up areas, cities, and regions, and the production of interstitial spaces that frames their role and characteristics within sprawling urbanisation. The agenda of research proposed in this book provides a more comprehensive understanding of the urbanisation phenomenon from its *non-built-up* dimension. The fact that the number of interstitial spaces of different kinds has been calculated to be substantial in cities – and may increase in contexts of urban shrinkage – also highlights the quantitative significance of interstices. However, the qualities of interstitial spaces are relevant considering their functions, spatial characteristics, relationality, politics involved around the urbanisation process, and the way how they contribute to the urban condition of cities. As reclaimed and contested spaces, interstices also contribute to the understandings of policies around urban growth, urban regeneration, and preservation of rural and environmental assets. The interstitial spaces also provide further streams of research around their contribution to the economy of cities, the politics associated with the production of the space, and the broad environmental role of interstices as elements of resilience in contexts of climate change.

1.4 The structure of the book

The book is organised in nine chapters. Chapter 1 begins with situating the interstitial spaces in debates about urbanisation and urban sprawl. Then the book follows a logical pathway – chapter by chapter – in which a conceptual framework around the notion of 'interstitial spaces' is developed and then the nature of interstitial spaces in relation to planning and the production of the space is discussed. The empirical section of the book introduces seven cases that illustrate the varied nature of interstitiality to finally discuss its implications in the broader field of urban studies.

Reflections upon further lines of enquiry and theories of urbanisation, urban sprawl, and cities are highlighted in the conclusion chapter. The organisation of the book is discussed in the following sections.

CHAPTER 1. Enquiring into urban sprawl: is it all about the built-up space?

Yet urbanisation can hardly be considered without reference to the interstitial spaces that lie between developments, cities, and regions; the book begins by highlighting the *interstitial space* as a relevant component of urban sprawl and as an object of study that deserves closer attention. It opens with references to the mainstream discussion on urban sprawl and how this debate has missed – or left behind – the interstitial dimension. This chapter argues the need of introducing a research agenda to unpack the somewhat vague interstitial geography of city-regions, problematised as the production of the space as a synonym of the *built-up*. This chapter situates the contrast between the built-up space and the interstitial geography and opens the discussion on the still unexplored role of interstices in the urbanisation process. The chapter states the aims, the research design, and the rationale of the book. This chapter also introduces the empirical base of the research – Santiago de Chile – as a suitable case study for in-depth examination of the interstitial spaces.

CHAPTER 2. The urban sprawl debate and the Zwischenstadt: where is the interstitial geography?

Chapter 2 addresses contemporary debates on urban sprawl and processes of extended urbanisation, while situating the emergence of interstitial geography. The chapter presents the debate on urban sprawl occurring within two approaches. First, urban sprawl relates to studies on traditional suburbanisation with a clear 'city-oriented' focus. The second approach relates to contemporary patterns of advanced suburbanisation in which metropolitan areas reach the status of city regions (Charmes & Keil, 2015; Phelps, 2018). Both approaches, however, acknowledge the interstitiality as part of complex spatial, political, and environmental processes of land fragmentation that characterise the sprawling developments of cities and regions. Thereby, the interstitial spaces and the geographical scope where they manifest are introduced: *inner suburban lands*, lands for *continuous urban expansion*, and the *urbanised countryside*. The latter includes wider rural geographies between cities and regions and frames the scope of interstitial spaces within debates of globalisation and planetary urbanisation. This chapter collates with the extant theories of the urban question in which interstitial spaces are part of, and argues for shifting the city-oriented focus in the studies of urbanisation.

CHAPTER 3. *The interstitial spaces of urban sprawl*

This chapter provides a comprehensive 'state of the art' by revising the extant literature that touches on such interstitial spaces in contexts of urban sprawl – including insights from architecture, urban planning, anthropology, geography, and environmental sciences – and processes of extended suburbanisation, urban fringe belts, and Sieverts's (2003) *Zwischenstadt*. There is a variety of terms that refer to different kinds of interstitial spaces: 'undeveloped space' (Theobald, 2001; Wolman et al., 2005), 'vacant lands' (Foo et al., 2013; Ige & Atanda, 2013; Northam, 1971), 'brownfields' (Pagano & Bowman, 2000), 'open spaces' (Barkasi et al., 2012; Kurz & Baudains, 2010), 'wildscapes' (Jorgensen & Keenan, 2012), 'wastelands' (Gandy, 2013), the 'drosscape' (Berger, 2006), 'non-urbanised areas (NUAs)' (La Rosa & Privitera, 2013), 'non-urbanised areas' (NUAs) (La Greca et al., 2011), 'non-places' (Augé, 1995), 'terrain vagues' (de Solá-Morales, 2002), 'urban voids' (Polidoro et al., 2011), 'interfragmentary space' (Vidal, 1999, 2002), 'inter-places' (Phelps, 2017), 'loose space' (Frank & Stevens, 2007), and 'lost space' (Trancik, 1986). These are all conceptual approaches that refer to particular types of interstices defined as more or less integrated leftovers; outcomes of less controlled processes in planning that are somehow claimed to be recovered, or as large geographies with economic potentials. The chapter also critically revises current uses of the term *interstice* in urban studies to finally develop an in-depth conceptual revision of the term 'interstitial space' applied to urban studies. This review places the conceptual versatility of the term 'interstitial space' and its practical effects in the analysis of cities and processes of sprawling urbanisation. This includes the infrastructural potential of interstices for establishing networks of unregulated spaces where ecological and socio-cultural diversity can flourish as part of the urban. The chapter finally develops insights into the 'interstitial geography': the wider context of interstitial spaces and hubs that emerge within the interstitial spaces.

CHAPTER 4. *The nature of interstitial spaces*

Chapter 4 explains how interstitial spaces emerge, how they are produced, what determines their presence, and what their significance is in planning practice. All of these aspects shape the nature of interstitial spaces as outcomes of the processes of urbanisation. It is argued that interstitial spaces are not mere leftovers of less controlled processes in planning. Instead, they are originated alongside the urbanisation process as unexpected outcomes of planning, while in many cases, the interstices are explicitly defined by planning decisions, economic constraints, or cultural ideals around the 'suburban dream' (Phelps, 2012). The origins of interstitiality are presented as an outcome of a combination of factors that amalgamate normative and discretional planning. The origins of interstitial spaces also determine their significance as elements that can be perceived as incidental gaps in the fabric of cities and regions, as well as opportunities for further developments and innovations in planning. This chapter highlights the inseparability of analytical perspectives and normative agendas on urban dynamics as much as the built environment.

CHAPTER 5. *The scales of the interstitiality*

In Chapter 5, it is argued that urban interstices cannot be reduced to their morphological aspects, as they are nevertheless spatially defined and distributed. Therefore, the analytical perspective that speaks to multiple geographical scales is the one that operationalise their analysis. Based on this, four scales at which the interstitial spaces can manifest are proposed: *proximity, transition, regional,* and *remoteness.* These scales are determined by the diverse composition of interstices, their surroundings, and their simultaneous place in both local and regional processes of urbanisation. A suburban river, for instance, can be understood as a narrow space across which people from one side can see the people on the other side, and thus, describe a spatial scale of 'proximity'. However, the same river can cross the entire city – while connecting different districts – and even extend beyond the city by reaching the regional space. Conversely, an open space can be spatially large but nevertheless part of a single administrative area. This is the case for a metropolitan park that has a large size while located within a single administrative area. Therefore, interstitial spaces can present a transcalar salience that ranges from the emergence of liminal spaces in suburban areas up to colossal landscapes between cities and regions. The notion of 'scale' has spatial, functional, and political dimensions that cover a wide range of local and regional processes of urbanisation, with implications in terms of governance and the way of how the interstitial spaces and cities build a mutual relation of influence.

CHAPTER 6. *The relational character of interstitial spaces*

Chapter 6 draws upon the relationality of interstitial spaces from spatial and ontological perspectives. It is proposed that the relationality of interstitial spaces is defined by three aspects: (a) *infrastructures*, (b) *functions*, and (c) *spatiality*; these aspects operate as interlinked while defining the relationality of interstices. In terms of infrastructure, spatial mobility and networks are discussed as critical elements that connect the interstitial spaces with local, regional, and global networks of the urbanisation process. Functions also influence the way of how interstitial spaces attract people, activities, uses, and infrastructures that connect the interstices with local or regional surroundings. Finally, spatial configurations illustrate how interstices operate as connectors or disruptors of urban processes and the city, in which borders, boundaries, and boundary objects play a critical role in defining the relationality of interstitial spaces. So, an interstitial space can be a zone of reconciliation, connection, integration, or a barrier that increases spatial segregation. These aspects of relationality are explained from both theoretical and empirical angles; one of them relates to Vidal's notion of urban fragmentation, which suggests that 'the urban phenomenon is essentially a permanent tension between fragments' (Vidal, 2002, p. 150); a tension that occurs through the interstitial space. In this vein, the interstitial space is a field of interaction, integration, or disconnection where transitions between fragments occur.

CHAPTER 7. Exploring seven interstitial spaces in Santiago de Chile

This chapter presents seven interstitial spaces in Santiago de Chile. This is the empirical section of the book and starts by describing the capital city of Chile – Santiago – and the interstices selected for analysis. The selected cases are mainly located in the southern metropolitan area of Santiago as this area has experienced the highest rate of suburban expansion over the last forty years; an expansion that describes embryonic patterns of polycentrism while spanning outer farming lands and rural districts, conurbations, and satellite towns. The seven cases represent the diverse nature of interstitiality, their origins, relation to planning, scale, and relationality, and are strategic for the city and the general urbanisation process. This chapter discloses the spatial, functional, and infrastructural character of these interstices and their meanings for different actors regarding planning opportunities. These selected interstitial spaces illustrate well the significance of non-urban geographies in contexts of urban sprawl that reaches regional scales (Figure 1.2).

The two conurbation zones are considered as a single category. The difference between them is that the first (Maipu/Padre Hurtado) comprises two independent communes, while the second (San Bernardo/Lo Herrera) encompasses two localities within the same administrative area. These conurbations are useful for understanding the implications of interstitiality in the governance of Santiago's sprawling growth.

CHAPTER 8. The implications of interstitial spaces in urban studies

Chapter 8 reflects upon the implications of interstitial spaces in urban studies. The interstitial spaces are active elements of the sprawling urbanisation of cities and regions and emerge as platforms of economic exchange, mobility, relationality, transition, and attraction of diverse interests around environmental preservation and change. The interstitial spaces encapsulate spatial relations within the sprawling development of cities and beyond, and thus, they have an integrative nature that can support urbanisation projects of different sorts ranging from urban regeneration up the transformation

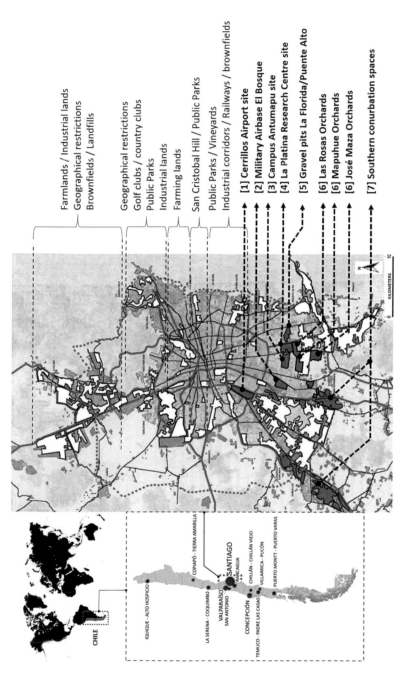

Farmlands / Industrial lands
Geographical restrictions
Brownfields / Landfills

Geographical restrictions
Golf clubs / country clubs
Public Parks
Industrial lands
Farming lands

San Cristobal Hill / Public Parks

Public Parks / Vineyards

Industrial corridors / Railways / brownfields

[1] Cerrillos Airport site
[2] Military Airbase El Bosque
[3] Campus Antumapu site
[4] La Platina Research Centre site

[5] Gravel pits La Florida/Puente Alto

[6] Las Rosas Orchards
[6] Mapuhue Orchards
[6] José Maza Orchards

[7] Southern conurbation spaces

Figure 1.2 Map of Santiago, its interstitial spaces and the selected cases for analysis.

of the regional space. The interstices can support the creation of planning schemes that are latent to the non-urban geographies between cities and offer alternative points on entries in the studies of suburbanisation, urban sprawl, and global processes of urbanisation. The nature of the elements that compose these non-urban geographies, along with the commodification and chains of collective consumption associated with interstitial spaces – contest the idea that 'cities' and 'the urban' are an undifferentiated object of study. Politics – as illustrated through the examples of Santiago de Chile – are never absent from the fortunes of interstitial spaces and have implications in the extant theories of urban politics.

CHAPTER 9. Conclusions

Chapter 9 presents conclusions and avenues for future research. In this chapter, the findings are reviewed in relation to the implications of interstitial spaces in urban theories and planning practice. It develops a series of theoretical reflections that reinforce the idea that sprawling developments are not only a matter of built-up areas but also interstitial spaces. Theoretically, the notion of 'interstitial space' appears more embracing than other existing terms and although rarely used in planning, it has some conceptual advantages to unify the analysis of the non-urban geographies of cities. Accordingly, Sieverts' (2003) *Zwischenstadt* appears as a context to be revisited from its 'non-city-oriented' dimension, as the fact that interstitial spaces emerge as constitutive of, and suggests a distinctive geography that deserves planning and policy approaches in its own right. As such, the 'interstitial spaces' of the *Zwischenstadt* encapsulate different possibilities of juxtapositions, hybridisation, intersections, embeddedness, articulations, uncertainty, and strangeness, and point towards more complex associations in urban governance and the politics associated to 'the urban'.

Interstitial spaces are active non-standard elements of the *Zwischenstadt*' and contest the ontologies about 'the urban'. Interstitial spaces are politically mediated – while apparently marginalised – and trigger the illusion of being reclaimed for the city. They are nevertheless spaces of resistance and preservation of socio-environmental, political, and economic values. The interstices entail the reimagination of suburbia and the expansion of urban fragmentation towards non-urban geographies. Their emptiness, infrastructural character, and spatial and ecological diversification force the adoption of interdisciplinary approaches for the study of the built environment. The interstitial spaces – irrespective of their differential visibility in the sprawling growth of cities and regions – are a realm left behind from the obsession to cities – even when some of the truly radical changes are currently taking place in the interstitial spaces. In that sense, the susceptibility of interstices to both scalar and relational ontologies suggests the development of theories of urban politics beyond 'the city' or 'the urban' as a single undifferentiated unit.

References

Abubakar, I. R., & Doan, P. L. (2017). Building new capital cities in Africa: Lessons for new satellite towns in developing countries. *African Studies*, *76*(4), 546–565.

Angelo, H., & Wachsmuth, D. (2015). Urbanizing urban political ecology: A critique of methodological cityism. *International Journal of Urban and Regional Research*, *39*(1), 16–27.

Arboleda, M. (2016). In the nature of the non-city: Expanded infrastructural networks and the political ecology of planetary urbanisation. *Antipode*, *48*(2), 233–251.

Augé, M. (1995). *Non-places. Introduction to an anthropology of supermodernity*. Verso.

Barkasi, A., Dadio, S., Losco, R., & Shuster, W. (2012). Urban soils and vacant land as stormwater resources. *World environmental and water resources Congress 2012*. American Society of Civil Engineers. 569–579.

Barnes, K. B., Morgan, J. III, Roberge, M., & Lowe, S. (2001). *Sprawl development: Its patterns, consequences, and measurement* (pp. 1–24). Towson University.

Barrington, C., & Millard, A. (2015). A century of sprawl in the United States. *Proceedings of the National Academy of Sciences*, *112*(27), 8244–8249.

Berger, A. (2006). *Drosscape: Wasting land urban America*. Princeton Architectural Press.

Borsdorf, A., Hidalgo, R., & Sanchez, R. (2007). A new model of urban development in Latin America: The gated communities and fenced cities in the metropolitan areas of Santiago de Chile and Valparaíso. *Cities*, *24*(5), 365–378.

Bowen, G. A. (2009). Document analysis as a qualitative research method. *Qualitative Research Journal*, *9*(2), 27–40.

Brenner, N., & Elden, S. (2009). Henri Lefebvre on state, space, territory. *International Political Sociology*, *3*, 353–377.

Brenner, N., & Schmid, C. (2014). The 'urban age' in question. *International Journal of Urban and Regional Research*, *38*(3), 731–755.

Bruegmann, R. (2005). *Sprawl: A compact history*. University of Chicago Press.

Burawoy, M. (1991). The extended case method. *Ethnography unbound. Power and resistance in the modern metropolis* (pp. 271–290). University of California Press.

Burger, M., & Meijers, E. (2012). Form follows function? Linking morphological and functional polycentricity. *Urban Studies*, *49*(5), 1127–1149.

Charmes, E., & Keil, R. (2015). The politics of post suburban densification in Canada and France. *International Journal of Urban and Regional Research*, *39*(3), 581–602.

Chen, G., Hadjikakou, M., Wiedmann, T., & Shi, L. (2018). Global warming impact of suburbanization: The case of Sydney. *Journal of Cleaner Production*, *172*, 287–301.

Collier, J., & Collier, M. (1986). *Visual anthropology: Photography as a research method*. UNM Press.

Da Costa Braga, A., & Rigatti, D. (2009). International conurbations along Brazil – Uruguay border on ambiguity drives spatial patterns and social exchange. In D. Koch, L. Marcus, & J. Steen (Eds.), *Proceedings of the 7th international space syntax symposium*. KTH.

de Solá-Morales, I. (2002). *Territorios [Territories]*. Editorial Gustavo Gili. S.A.

Ernstson, H. (2021). Ecosystems and urbanization: A colossal meeting of giant complexities. *Ambio*, *50*, 1639–1643.

Evans, C., & Lewis, J. (2018). *Analysing semi-structured interviews using thematic analysis: Exploring voluntary civic participation among adults*. SAGE Publications Limited.

Fishman, R. (1987). *Bourgeois utopias: The rise and fall of suburbia.* Basic Books.

Flyvbjerg, B. (2006). Five misunderstandings about case-study research. *Qualitative Inquiry*, *12*(2), 219–245.

Foo, K., Martin, D., Wool, C., & Polsky, C. (2013). The production of urban vacant land: Relational placemaking in Boston, MA neighborhoods. *Cities*, *35*, 156–163.

Frank, A., & Stevens, Q. (2007). *Loose space. Possibility and diversity in urban life.* Routledge.

Gallent, N., & Shaw, D. (2007). Spatial planning, area action plans and the rural-urban fringe. *Journal of Environmental Planning and Management*, *50*(5), 617–638.

Galletta, A. (2013). *Mastering the semi-structured interview and beyond: From research design to analysis and publication* (Vol. 18). NYU Press.

Gandy, M. (2013). Marginalia: Aesthetics, ecology, and urban wastelands. *Annals of the Association of American Geographers*, *103*(6), 1301–1316.

Gavin, H. (2008). Thematic analysis. In H. Gaving (Ed.), *Understanding research methods and statistics in psychology* (pp. 273–281). SAGE Publications Limited.

Harrison, J., & Heley, J. (2015). Governing beyond the metropolis: Placing the rural in city-region development. *Urban Studies*, *52*(6), 1113–1133.

Heynen, N., Kaika, M., & Swyngedouw, E. (2006). Urban political ecology. In N. Heynen, M. Kaika and E. Swyngedouw (Eds.), *The nature of cities: Urban political ecology and the politics of urban metabolism* (pp. 1–20). Routledge.

Ige, J. O., & Atanda, T. A. (2013). Urban vacant land and spatial chaos in Ogbomoso North local government, Oyo State, Nigeria. *Global Journal of Human Social Science & Environmental Science & Disaster Management*, *13*(2), 28–36.

Jorgensen, A., & Keenan, R. (2012). *Urban wildscapes.* Routledge.

Kassa, F. (2013). Conurbation and urban sprawl in Africa: The case of the city of Addis Ababa. *Ghana Journal of Geography*, *5*(1), 73–89.

Keil, R. (2017). *Suburban planet: Making the world urban from the outside in.* Polity.

Kennedy, M., Butt, A., & Amati, M. (2016). *Conflict and change in Australia's peri-urban landscapes.* Routledge.

Kropf, K. (2011). Morphological investigations: Cutting into the substance of urban form. *Built Environment*, *37*(4), 393–408.

Kurz, T., & Baudains, C. (2010). Biodiversity in the front yard: An investigation of landscape. Preference in a domestic urban context. *Environment and Behavior*, *44*(2), 166–196.

La Greca, P. L., Rosa, D., Martinico, F., & Privitera, R. (2011). Agricultural and green infrastructures: The role of non-urbanized areas for eco-sustainable planning in metropolitan region. *Environmental Pollution*, *159*(01), 2193–2202.

La Rosa, D., & Privitera, R. (2013). Characterization of non-urbanized areas for land-use planning of agricultural and green infrastructure in urban contexts. *Landscape and Urban Planning*, *109*(01), 94–106.

Lefebvre, H. (1970 [2003]). *The urban revolution.* University of Minnesota Press.

Mazzoni, C., & Grigorovschi, A. (2015). Strasbourg Eurométropole, a cross-border conurbation towards new sustainable mobility patterns. *Spatium*, *33*, 18–25.

Monkkonen, P., Comandon, A., Escamilla, J. A. M., & Guerra, E. (2018). Urban sprawl and the growing geographic scale of segregation in Mexico, 1990–2010. *Habitat International*, *73*, 89–95.

Northam, R. (1971). Vacant land in the American City. *Lands Economics*, *47*(4), 345–355.

Pagano, M., & Bowman, A. (2000). Vacant land in cities: An urban resource. *The Brookings Institution – Survey series* (pp. 1–9). Center on Urban & Metropolitan Policy.

Phelps, N. (2012). *An anatomy of sprawl. Planning and politics in Britain.* Routledge.

Phelps, N. A. (2015). *Sequel to suburbia: Glimpses of America's post-suburban future.* MIT Press.

Phelps, N. A. (2017). *Interplaces: An economic geography of the inter-urban and international economies.* Oxford University Press.

Phelps, N. (2018). In what sense a post-suburban era? In B. Hanlon, & T. Vicino (Eds.), In *The Routledge companion to the suburbs* (pp. 39–47). Routledge.

Phelps, N., & Wood, A. (2011). The new post-suburban politics? *Urban Studies, 48*(12), 2591–2610.

Polidoro, M., de Lollo, J. A., & Barros, M. V. F. (2011). Environmental impacts of urban sprawl in Londrina, Paraná, Brazil. *Journal of Urban and Environmental Engineering, 5*(2), 73–83.

Rayback, S. (2016). Making observations and measurements in the field. In N. Clifford, M. Cope, S. French, & T. Gillespie (Eds.), *Key methods in geography* (pp. 325–335). Sage.

Roberts, L. (2016). Interpreting the visual. In N. Clifford, M. Cope, S. French, & T. Gillespie (Eds.), *Key methods in geography* (pp. 233–247). Sage.

Scott, A. J., & Storper, M. (2014). The nature of cities: The scope and limits of urban theory. *International Journal of Urban and Regional Research, 39*(1), 1–15.

Shaw, K. (2015). Planetary urbanisation: What does it matter for politics or practice? *Planning Theory & Practice, 16*(4), 588–593.

Shaw, P., & Hudson, J. (2009). The qualities of informal space: (Re) appropriation within the informal, interstitial spaces of the city. In *Proceedings of the conference occupation: Negotiations with constructed space* (pp. 1–13). University of Brighton.

Sieverts, T. (2003). *Cities without cities. An interpretation of the Zwischenstadt.* Spon Press.

Sousa Matos, R. (2009). Urban landscape: Interstitial spaces. *Landscape Review, 13*(1), 61–71.

Stender, M. (2017). Towards an architectural anthropology – What architects can learn from anthropology and vice versa. *Architectural Theory Review, 21*(1), 27–43.

Storper, M., & Scott, A. J. (2016). Current debates in urban theory: A critical assessment. *Urban Studies, 53*(6), 1114–1136.

Tashakkori, A., & Creswell, J. W. (2007). Exploring the nature of research questions in mixed methods research. *Journal of Mixed Methods Research, 1*(3), 207–211.

Theobald, D. M. (2001). Land use dynamics beyond the American urban fringe. *Geographical Review, 91*(3), 544–564.

Trancik, R. (1986). *Finding lost space; Theories of urban design.* Van Nostrand Reinhold Company.

Tschumi, B. (1996). *Architecture and disjunction.* MIT Press.

UN (2018). *Revision of world urbanization prospects.* Department of Economic and Social Affairs, Population Division.

Vergara, F., & Boano, C. (2020). Exploring the contradiction in the ethos of urban practitioners under neoliberalism: A case study of housing production in Chile. *Journal of Planning Education and Research*, 1–15.

Vidal, R. (1999). Fragmentos en tensión: Elementos para una teoría de la fragmentación urbana [Fragments in tension: Elements for a theory of urban fragmentation]. *Revista Geográfica de Valparaíso* 29(30), 149–180.

Vidal, R. (2002) *Fragmentation de la Ville et nouveaux modes de composition urbaine [The fragmentation of the city and new modes of urban composition]*. Editions L'Harmattan.

Wolman, H., Galster, G., Hanson, R., Ratcliffe, M., Furdell, K., & Sarzynski, A. (2005). The fundamental challenge in measuring sprawl: Which land should be considered? *The Professional Geographer*, *57*(1), 94–105.

Yin, R. K. (2009). *Case study research: Design and methods*. Sage.

Zook, M., & Sheltom, T. (2013). The integration of virtual flows into material movements within the global economy. In P. Hall, & M. Hesse (Eds.), *Cities, regions and flows* (pp. 42–57). Routledge.

2 The urban sprawl debate and the *Zwischenstadt*

Where is the interstitial geography?

> There are many more houses in the valley than the small number of people who still farm there could possibly occupy, and it is possible to make out through the dense tree cover of the hillsides many other houses that clearly have no connection with agricultural production. This is not at all the completely rural scene that It might appear to be.
>
> (Bruegmann, 2005, p. 1)

2.1 Introduction

The term 'urban sprawl' appeared in the 1940s to describe the extensive American suburbanisation driven by the increasing use of private cars and the expansion of the interstate motorway system (Gillham, 2002; Soule, 2006). Since then, a longstanding stream of research and a vast body of literature has been established with incredible insights about the multifaceted character of urban sprawl – including the politics and ideological meanings associated with the type of suburbia that it suggests – and its transnational salience as urban sprawl is possible to be seen in most city-regions (Bruegmann, 2005; Couch and Karecha, 2006; Ewing et al., 2002; Galster et al., 2001; Johnson, 2001; Phelps & Wood, 2011; Zhang et al., 2013). Urban sprawl is a contradictory phenomenon in terms of politics, as while it can be credited with fostering global economic growth and stability it has also been pointed as the cause of significant repercussions on climate change and oil depletion (Phelps, 2015).

It is important to notice that after all, urban sprawl carries a (well-established) negative connotation considering its (well-documented) impacts – so, this chapter is not a plea in favour of – but a more distanced discussion around the concept itself, and the types of environment through which interstitial spaces are manifested. There is enough evidence to suggest that sprawling contexts encompass fundamental components of the urbanisation process – including the interstices that appear alongside the production of the space – although still play a secondary role to explain the urban condition in comparison to cities' centres: 'the fascination of the myth of the

DOI: 10.4324/9780429320019-2

Old City clouds our view of the reality of the periphery' (Sieverts, 2003, p. 12). Yet urban sprawl deserves a closer inspection to define more balanced narratives around its composition and explore what other elements can contribute to a more comprehensive understanding of urban sprawl.

Since its inception – and although the definition is still contested – 'urban sprawl' is used to describe extended suburban landscapes largely characterised by low-density residential neighbourhoods, high levels of car-dependency, single land uses, and lack of physical continuity (Jaret et al., 2009). Land fragmentation and dispersion of infrastructures are also components of sprawling growth (Altieri et al., 2014; Trubka et al., 2010). In terms of geographical delimitation, it is also accepted that urban sprawl embraces developments beyond suburbia, in which a range of terms have been used to identify its scope including 'exurbia' (Bruegmann, 2005), 'technoburbs' (Fishman, 1987), 'edge cities' (Garreau, 1991), 'edgeless cities' (Lang, 2003), 'satellite towns' (Abubakar & Doan, 2017) and 'post-suburbs' (Phelps & Wood, 2011); all constitutive of Sieverts' *Zwischenstadt*: a distinctive geography of transition between the proper city and the open countryside that amalgamates the characteristics and dynamics of both. The *Zwischenstadt* is a realm 'which separates itself from the city (...) and achieves a unique form of independence' (Sieverts, 2003, p. 6). This multifaceted geography has ramifications across several scales, notably in terms of governance. This issue has been identified by Rauws and de Roo's notion of the 'third type of landscape' (Sieverts, 2011), something 'which cannot solely be understood in terms of progressive intensification of urban functions in the rural environment' (Rauws & de Roo, 2011, p. 269), and 'be distinguished in terms of its own specific dynamics and characteristics (...), a spatial system in its own right' (ibid, p. 270). This distinction becomes relevant while stepping on the ontologies around 'urban sprawl': does it refer to 'the sprawl of *the urban*' or 'the sprawl of *the city*'? As an 'urban' phenomenon, the extension of its scope is not a problematic exercise considering that although cities can eventually have recognisable boundaries of some sort, 'the urban' is increasingly gaining traction in debates of planetary urbanisation.

There is enough evidence to indicate that rather than being natural or spontaneous (as many still suggest), urban sprawl has been the unintended consequence of underlying politics of modern capitalism (Beck et al., 2003); a pattern of urban development that 'has been thoroughly planned as a result of interventions by all ties of government' (Phelps, 2015, p. 40). These debates on urban sprawl have been framed by two epistemological approaches. First, a morphological line of enquiry in which urban design has become the champion against urban sprawl and second, a policy-based approach that somehow vindicates the importance of spatial planning as the vehicle for sustainable development. In this chapter, these approaches are discussed to acknowledge the presence of interstices as elements produced alongside the sprawling growth of cities and regions. The chapter also situates the emergence of interstitial spaces regarding the geographical scope where they manifest; these are (a) the *inner suburbia*, (b) the *contiguous expansion*, and (c) the *urbanised countryside*. The latter suggests that sprawling growth – and the interstices – can reach endless magnitudes, an extension that talks about the hyper-regional scope of both the urban and the interstitial conditions.

2.2 Traditional sprawl and the morphological approach

First, urban sprawl is often associated with theories of the urban form with a clear morphological focus. On this basis, urban sprawl is depicted as a dispersed distribution of low-density neighbourhoods, infrastructures, large-scale architectonic artefacts – such as airports, energy plants, shopping malls, manufacturing industries, and others – all connected by roads and interspersed with undeveloped areas and open tracts of different sorts (Lang, 2003).

If we take the 'urban form-oriented' view to urban sprawl, then the sprawling landscape is the result of 'what has been done' – which is the object of study of morphological theories – rather than 'what is left in between'. This supposes a fundamental constraint in an 'urban' morphology approach while trying to capture the nature of the rural space or the 'unbuilt'. In morphological terms, the assumption that urbanisation is a derivation of planning the *built-up* entails that open tracts are somehow static spaces – eventually, a matter of land subdivisions – that sooner or later would become urbanised. That is why the politics associated with farming spaces placed in sprawling morphologies talk about their increasing functional decline while raising narratives of preservation and change (Roche & Argent, 2015; Silva, 2019). The encroachment of the rural space has accumulated an important body of research. There is abundant literature about the economic, social, and infrastructural impacts of urban sprawl on rural lands; literature that has fuelled anti-sprawl narratives of different sorts.

Most of these debates have derived into normative restrictions to urban growth such 'green belts' (Dockerill & Sturzaker, 2019), 'urban limits', and 'urban boundaries' (Lang, 2002; Morrison, 2010). However, these morphology-based measures are increasingly discredited as they do not restrain urban sprawl at all and even work in the opposite way resulting in greater leapfrogged urbanisation of the countryside (MacLachlan et al., 2017) and distortions in the peri-urban land market (Balducci, 2017; Branch, 2018). The morphological view on urban sprawl has meant the constraints of its policies of control: the London Plan, like many anti-sprawl plans since, was based on a simple and static view of the proper shape of the city' (Bruegmann, 2005, p. 176). Alternatively, narratives of retrofitting and compactness have emerged as more strategic when compared with normative restrictions to urban growth. These alternatives 'form something of a *Zeigeist* in architectural, design and planning circles'

(Phelps, 2015, p. 43) as they work along with 'the rise of new planning mantras such as "sustainable", "compact" or "zerocarbon" development' (Phelps, 2012, p. 172). Processes of retrofitting suburbia have been part of strategies in which densification plays a key role in configuring a more 'compact sprawl' (Rice, 2010; Ståhle & Marcus, 2008).

These two planning reactions – restrictions and retrofitting – have somehow overlooked the literature around the underlying determinants of urban sprawl that indicates that this is much a product of modernists-inspired nations where cities reflect projects of national development. This modernisation is driven by the values of the neoliberal agenda that has reached trans-national implications since these values 'have become embedded in the politics and praxis of governments, institutions and organisations – at all levels of spatial governance – around the globe' (Boland et al., 2017). The 'urban form-oriented' approach aims to reconvert vacant lands as its ultimate achievements connect to the invisibilisation of urban sprawl – considering its negative connotations – along with the implementation of flagship projects and urban densification (Ponzini, 2014; Tarazona Vento, 2017). These elements are recognised as part of the commodification of cities by having a place in world rankings of liveability, safety, sustainability, happiness, and others (Insch, 2018). In this city branding exercise, architecture and urban design become the primary platforms to sustain innovations in urban regeneration, and create urban landscapes with international salience (Knox, 2018). These are the cases of cities in Australia, New Zealand, Canada, the Middle East – and other neoliberalised nation-states – which compete to become attractive places for high-skilled migrants, foreign investments, and tourists (Insch & Walters, 2018; Kearns & Lewis, 2019) while urban sprawl and its associated issues of housing affordability, poor public transport, and socio-spatial segregation go rampant all over the agriculturally fertile hinterland and culturally significant peri-urban landscapes (Curran-Cournane et al., 2016; Silva, 2019). In this context, open tracts are spaces to be urbanised as their condition as undeveloped is perceived as a waste of resources (Kim et al., 2018).

Technical studies on the 'sprawl index' identify patterns of urban sprawl by combining different indicators that change in accordance with urban growth dynamics (Wolman et al., 2005). Although part of the realm of morphological studies, measuring the sprawl index supposes that urban sprawl is not a static morphological picture of fragmented suburbia but a dynamic 'pattern of land use in an urbanised area that exhibits low levels of some combination of eight distinct dimensions: density, continuity, concentration, clustering, centrality, nuclearity, mixed uses, and proximity' (Galster et al., 2001, p. 685). If we take Galster's process-oriented view on urban sprawl, then the sprawling landscape is not unique and homogeneous but diverse and changing as it evolves over time and can differ from region to region. Indeed, different areas of the same city can also show different levels of sprawling growth (Li et al., 2020): 'it is the trend in population density, rather than current population density, that determines whether a city is sprawling or not. A city becoming less densely populated through time is said to be sprawling, even if it is currently quite densely populated in comparison to other cities' (Hess et al., 2001, p. 6). On this basis, urban sprawl is a matter of 'degree' that must be accepted as 'a continuous process of urban transformation which functions more like a verb than a noun' (Silva, 2019, p. 58); as such, it can also be observed in small towns and villages. This makes urban sprawl a non-unique condition of megacities (Horn & Van Eeden, 2018), which raises further reflections around its intrinsic unsustainable character. Other technical studies have addressed urban sprawl through its patterns of land occupation – trying to find out the underlying regularities behind stable variables of material and spatial change – which 'have exited the interest of "fractal researchers" who have attempted to represent these growth processes in mathematical terms, with extremely interesting results' (Sieverts, 2003, p. 7).

2.3 The evolution of urban sprawl and its underlying politics

Some argue that morphological approaches are insufficient to understand and manage urban sprawl as this is the outcome of less controlled processes in planning (Talen, 2010). Debates around policy-based determinants have identified land-use conversion, population change, traffic and vehicle miles travelled, energy consumption, and fiscal measures – such as impact fees and taxation on outer developments (Nelson, 1999) – as important policy factors. However – and setting aside the contextual and historical differences – there is a consensus that since the 1940s, governments of all kinds have articulated a national interest in the revitalisation of central areas while promoting the massive redistribution of population and infrastructure across the suburbs (Phelps, 2015). In general terms, major investments in road infrastructure, incentives in the form of mortgage relief, and permissive planning controls of rural counties stimulated development on unincorporated land. More specifically, empirical studies conducted in China (Li & Li, 2019; You & Yang, 2017), North America (Barrington & Millard, 2015), Europe (Oueslati et al., 2015; Pirotte & Madre, 2011), Latin America (Silva and Vergara-Perucich, 2021), and Middle East (Bagheri & Tousi, 2018; Masoumi et al., 2018) demonstrate that determinants of urban sprawl – although varied – are possible to be sorted into physical, geographical, socioeconomic, and policy-based determinants.

In European cities, urban sprawl is influenced by policies on population growth, housing, land market constraints, and improvements on infrastructure (Oueslati et al., 2015; Pirotte and Madre, 2011). It is important to notice that these studies are limited to monocentric city models with clear correlations between land fragmentation and income growth (that derives into higher rates of car use). In the case of China, driving forces of urban sprawl differ across spatial scales linked to the hierarchical governance, and the socioeconomic and political factors that intensify people's mobility between rural, urban, and intra-urban areas (Li et al., 2020). In some European cities, upwards mobility and economic security – along with the perception of a 'good living' associated with detached or semi-detached homes with private gardens in a rural or suburban living environment – reinforce the 'suburban dream' and enhance location on peri-urban areas (Bontje, 2004). Similarly, there is a correlation between perceived rates of crime and land affordability that encourage suburban locations (Qian & Wong, 2012). An extensive study carried out in Switzerland at the municipal level found that determinants of urban sprawl are linked to prices of agricultural lands, value of land for construction, population growth, land accessibility, federal tax, changes in employment growth, number of homeowners, income and commuting rates by private and public transport (Weilenmann et al., 2017). Studies conducted in Lagos, Nigeria, identified spatial policies as the main determinants of urban sprawl, influencing the characteristics of neighbourhoods, proximity to water bodies, and economic growth. The skewed distribution of private land, the high costs of undeveloped land, and flexible regulations also influence sprawling growth alongside accessibility and land-use change of forests and farmlands (Braimoh & Onishi, 2007). In Latin America, aggressive housing policies, dispersion of infrastructure, low prices of rural lands, the emergence of peripheral slums – all framed by strong neoliberal policies in planning – appear as clear elements of sprawling suburbanisation (Borsdorf et al., 2007; Silva, 2019; Tapia 2013; Vergara and Boano, 2020). Policies of functional intensification – as a homologue of retrofitting – have also been consistent in the management of urban sprawl and have influenced the emergence of more multifunctional suburban environments that accommodate other than solely residential land uses: 'since their creation, the suburbs have been evolving and changing. From bedroom communities to edge cities, the trend has been towards more complex and complete places' (Calthorpe & Fulton, 2001, p. 198).

Empirical studies on current processes of land-use change in the peri-urban space of city-regions found that sprawling growth – still characterised by spatial fragmentation – illustrates emerging patterns of polycentricity and functional self-sufficiency in economic, social, and political terms (Huang et al., 2016; Burger & Meijers, 2012). This is particularly relevant, as polycentric city-regions poses important challenges to anti-sprawl narratives considering embedded elements of urban sustainability based on the reinterpretation of the functional linkage between work and residence (Phelps & Wood, 2011). Polycentric city-regions depict patterns of sprawling growth in which suburban sub-centres articulate not only centre-periphery relations but also periphery-periphery linkages. It is argued that 'polycentric development is more efficient and sustainable than compact and monocentric development when an urban area achieves a certain size, with reasonable proximity and functional independence between centres' (Qian & Wong, 2012, p. 405). The sprawling space of polycentric cities becomes multifunctional, and urban fringes are complex environments because of their random assemblage of shapes, former industrial functions, polluted lands, geographical accidents, bulk-retail, degraded farmlands, large recreational areas and open tracts, fragmented residential developments, and transitional lands for trips and temporary activities. This complexity – in which undeveloped lands are part of – defines a diverse space of transformation, experimentations in planning, and alternative policy approaches beyond bi-dimensional conceptions of land-use (Gallent & Shaw, 2007).

In advanced stages of polycentrism, functionally independent urbanised areas concentrate industries with residential and commercial land-uses. These areas tend to become more self-contained and, in some cases, take the form of suburban 'minicities' that configure constellations of interconnected 'peri-centres' (Salet & Woltjer, 2009). Based on how Oslo has defined its sprawling space, Røe and Saglie (2011) propose a model for urban sustainability through the consolidation of these suburban 'minicities' in the form of interconnected districts characterised by a mix of residential land-uses, workplaces, housing typologies, public amenities, transportation and public spaces with comparatively higher indicators of urban quality (Røe & Saglie, 2011). The notion of 'minicity' finally correlates with closely related literature on post-suburbansation (Phelps et al., 2010; Phelps, 2015; Phelps, 2018; Phelps & Wood, 2011). Aside from the morphological features and the location of edge-urban developments, the diverse sprawling space of city-regions evinces that some peri-urban agglomerations have acquired specific functions and become increasingly independent from the traditional city. As Heinrichs et al. (2011) noted, 'the observation that the rise of edge cities, technoburbs and the like has largely led to a disconnection of fringe developments from their former point of reference in the centre of the old cities is one of the pillars of the concept of post-suburbia' (p. 107). Compared with suburbia, '...the most notable functional difference is the more balanced employment and residential character of post-suburbia' (Phelps, 2015, p. 32). It is also important to notice that post-suburbanisation would refer to the kind of urbanisation that is taking place beyond the formerly suburbanised area, in the still rural hinterland. This talks about the geographical scale of post-suburbanisation in which 'post-suburban communities and their politics can and should be positioned within wider metropolitan urban systems. That is, post-suburbs take their place among a range of different settlement types across metro regions' (ibid, pp. 8–9).

In some way, these ideas were entailed by Sieverts (2011) when asserting that the *Zwischenstadt* can become 'a more open settlement between the city and country, which develops further with its own workplaces and facilities into a more or less independent *Zwischenstadt*...only then...it frees itself from its original dependency, supplies itself and enters into a relationship of mutual exchanges with the original city' (Sieverts, 2003, p. 4). For Sieverts, the evolution of this interdependency will finally depict a more clustered environment of (more or less) consolidated outer agglomerations, in which 'the city of tomorrow will consist of a concentration of compact settlements surrounded and surrounding countryside, which meets specific functions' (ibid, p. 43). In this new scenario, the city and the countryside will be amalgamated for the construction of a new 'urban', where a new symbiosis will define a more hybrid landscape of biotechnical systems in the city and 'new wilderness in the countryside' (ibid, p. 43). What is proven by Sieverts' conceptualisation of the *Zwischenstadt* is that its homologous American counterpart – 'urban sprawl' – has often referred (maybe accidentally) to 'the sprawl of the city' rather than 'the urban'; though we do not much say 'city's sprawl' – an entity we somehow have the conceptual apparatus to be analysed. This clarification is fundamental when locating the interstitial spaces as part of 'the urban' and thus, 'urban sprawl'.

2.4 Introducing the interstitial space

The evolution of urban sprawl from purely suburban arrangements towards more polycentric and post-suburban developments depicts the regional space of a continuum land-use change that involves both the production of built-up lands and undeveloped areas that lie between developments. If urban sprawl has been left behind the usual readings of urban and theories about the city, the interstitial spaces that lie between developments are definitely almost invisible – receiving little attention regarding their role in processes of extended (sub)urbanisation. Partly because of the existence of these interstices and their regional scale, the fragmented urban landscape is not yet seen as part of our culture (Sieverts, 2011). While the emphasis of Sieverts' discussion around the (in-between) *Zwischenstadt* is not upon *interstitial spaces* per se, one of the key points relates to the significance of fragments of undeveloped land that exist at the regional scale. The *Zwischenstadt* is not a continuous built-up realm but an 'urbanised countryside' – a geographical space that is 'neither city nor landscape, but which has characteristics of both' (Sieverts, 2003, p. 3). This suggests that undeveloped areas and the open landscape operate as active components of the urban sprawl context. Sieverts argues that 'the grain and density of development of the individual urban areas and the degree of penetration with open spaces and landscapes determine the specific character of the *Zwischenstadt*', in which 'the landscape becomes the glue of' (ibid, p. 9). Thus, the *Zwischenstadt* depicts the whole suburban context as a scattered process of development where built-up lands coexist with various pieces of open countryside, forest and woodlands, open tracts, farming areas, natural handicaps, protected ecological reservoirs, flood valleys, restriction zones, conurbation zones, regional parks, green corridors, security

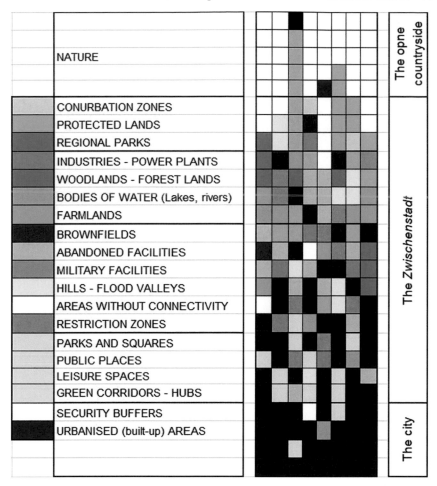

Figure 2.1 The spectrum of undeveloped areas and open tracts of the *Zwischenstadt*.
Source: Author.

buffers, industrial sites, and others that define a varied and functionally diverse geography (Figure 2.1).

The *Zwischenstadt* is characterised by different degrees of urbanisation and its diversity relates to the extent and scale of undeveloped land. In the *Zwischenstadt*, as others have described, open areas are a result of the mixing between social and economic demands from urban surroundings (Tacoli, 1998; Wang et al., 2012). Here, the countryside shows some expressions of urbanity – such as farmlands for public visits, tourism activities, and leisure places – and in which 'as a tool of experience and sign of identity, public space is more important than ever for the intelligibility and legibility of the *Zwischenstadt'* (Sieverts, 2003, p. 23). In this context, 'the distinction between "original" nature and "technically manipulated" nature becomes more and more difficult' (ibid, p. 39). For Sieverts, the interstitial spaces have significance beyond their mere condition as natural (or random) leftovers of the urbanisation process. Urbanisation itself refers to a form of intervention that although focused on the built-up, it nevertheless shapes all aspects of the built environment including its resulting interstitial geography: 'the open countryside has become an internal figure against the "background" of the settled area. The settled area itself could be interpreted as a special form of "landscape", which contains the open land. Both areas, the internal landscape free of development and the settlement itself are early completely man-made' (ibid, p. 40).

The land fragmentation that characterises the sprawling growth also confirms the presence of an interstitial geography. This fragmentation talks about the ambiguous urban-rural nature of sprawl through these non-urbanised lands confined by, within, or in-between urbanised fragments. Hebbert (1986) observed that 'the built-up area of any modern city tends to be surrounded by a transitional zone which is neither fully urban nor fully rural but instead displays a mixture of uses and building types interspersed with agricultural and vacant land' (p. 141). This ambiguity, however, is not such when these vacant lands are examined from outside the urban scope. What is an 'ambiguous' or 'undeveloped' or simply a 'gap' in the fabric of cities and regions is instead an 'ecotone' for environmental sciences; a very well-defined zone of transition between two or more different ecological communities (Odum & Barret, 2006). Somehow, ignoring the ecological (and other) values of interstitial spaces relates to the interpretation of urban sprawl and its physical fragmentation as a less efficient way of promoting urban development; a random process in which 'the result is a haphazard patchwork, widely spread apart and seeming to consume far more land than contiguous developments. Unless preserved or unbuildable, the remaining open tracts are usually filled with new developments as time progresses' (Gillham, 2002, pp. 4–5). This has important implications in looking at urban sprawl from an interdisciplinary perspective as it offers to structure the *Zwischenstadt*.

This transitional nature of the interstitial spaces and their links with the urbanisation process open questions about the geographic scope of the interstices, considering that Sieverts' *Zwischenstadt* embraces from the edges of the consolidated city up to the open countryside. The boundaries of this geographies space, however, are difficult to determine. There is some generic consensus, however, that 'we should explicitly refer to the kind of urbanisation that is taking place beyond the formerly suburbanized area, in the still rural hinterland' (Brake et al., 2005, cited in Phelps, 2015, p. 23).

2.5 The geographical scope of the interstitial spaces

Situating the geographical scope of interstitial spaces depends on the geographical scope of urban sprawl itself. It is clear that the sprawling expansion of cities has created metropolitan regions – largely formed by the concentration of different types of agglomerations – where residential land-uses are only one component of a varied geography that does not apparently have clear morphological and functional boundaries. This becomes more abstract when we consider the expansion of functional links supported by connectivity infrastructures that relate suburban areas with employment hubs located in the outer space of cities.

Attempts to define the geographical scope of urban sprawl are reasonably more conceptual as any approach to fit its spatial boundaries has been proved debatable or at least contradictory. The difficulties in identifying the geographical boundaries of sprawling areas, however, undermine the effectiveness of plans as boundaries differ in spatial, morphological, and administrative terms (Wolman et al., 2005). Morphologically speaking, a sprawling area is defined by patterns of dispersed suburbanisation driven by the increasing use of private cars, but this is confusing considering that transport infrastructure expands the sprawling hinterland (Gillham, 2002). This results in an un-boundaried urban sprawl that behaves more as a transitional geography in a state of constant change (Hess et al., 2001). Situating these boundaries, however, is necessary to understand the place and implications of interstitial spaces in the sprawling (sub)urbanisation process.

Aside from Sieverts' *Zwischenstadt*, Leontidou and Couch (2007) identified urban sprawl as comprehended somewhere between the cityscape, suburbs, and satellite towns, while van Leeuwen and Nijkamp (2006) indicates that urban sprawl is the geography of mutual urban-rural interactions between the town and its hinterland. This geography is formed by the town (that concentrates residences, industries, and services), the direct hinterland (that concentrates residences, labour, families, and cultural facilities), and the hinterland (that concentrates agriculture, regional transport, tourism, dispersed residential land uses and conservation areas). In a similar vein, Bradbury and Bogunovich (2013) indicate that Auckland's urban sprawl has a very detailed geography – without a clear rural-urban divide – defined by four zones: urban, suburban, peri-urban, and exurban. In the case of Santiago de Chile, Silva and Vergara (2021) identify at least four stages of urban development: the historical city, the consolidated city, the suburban, and the exurban. This is a case in which official policy reports clearly signalise a metropolitan main ring-road that circumscribes the entire city as the boundary between the city and its suburban districts (ibid). In that case, the suburban and exurban spaces would constitute the geography of urban sprawl.

More importantly is the fact that as well as the built-up spaces are dispersed all over the sprawling landscape, so, the interstitial spaces. In that sense, the spatial distribution of interstices defines different relations with their immediate surroundings and the urbanised areas that lie beyond. Also – and apart from its own spatial and functional characteristics – the surrounding landscape influences the spatial character of interstitial spaces. A large mining site, for instance, will be functionally influenced by the restrictions imposed by its immediate surroundings – whether this is open countryside, a protected ecological reservoir or suburban residential areas – and the regional space of supply and transportation of goods. Fringe belt areas will influence the rural character of interstices as well as inner lands will influence their suburban identity. Ultimately, this space would be virtually linked to the city centre and eventually other cities far beyond its immediate hinterland. Precisely, the lack of attention on this space of linkage confirms the underlying assumption that cities, their sprawling character and the urban condition rest solely on the effects of the built-up space rather than the interstices produced alongside the urbanisation process (Phelps & Silva, 2017). On this basis, the spatial composition of the sprawling urbanisation would help in determining the spatial position of interstices as part of.

The composition of the sprawling expansion differs from region to region, and from monocentric to polycentric modes of urban development. In a monocentric city, we would identify the different concentric rings of expansion in which the interstitial spaces are manifested. However, in the Randstad zone in the Netherlands – for instance – the polycentric development impedes situating the interstices as spaces initially linked the city's core, then the surrounding suburban space, then the peri-urban, etc. It does not impede, however, indicating that irrespective of functional and morphological patterns, the sprawling development embraces at least three geographical scopes at which the interstices are manifested: (a) *inner suburbia*, (b) *contiguous expansion*, and (c) the *urbanised countryside* (Figure 2.2).

2.5.1 The inner suburbia

Inner suburban interstitial lands are those that appear adjacent to consolidated suburban built-up areas. These interstices used to be located after (or closer) inner suburban zones of expansion. Their location and proximity to infrastructure and services make them attractive for further urbanisation, above all when these are abandoned, derelict, or in functional obsolescence. In other cases, these interstices have clear land uses as public space (squares, parks, or others) or functional uses that support infrastructure of connectivity. This is the case of railway lines between neighbourhoods, a motorway, the site of a power, or water plant that has been surrounded by the suburban expansion, or the buffer of security that

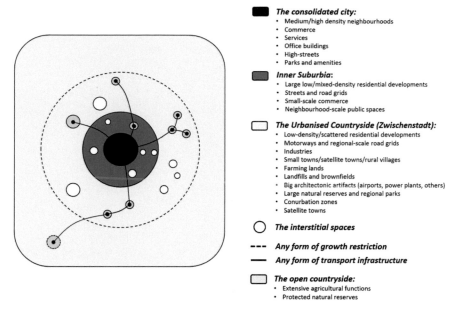

The consolidated city:
- Medium/high density neighbourhoods
- Commerce
- Services
- Office buildings
- High-streets
- Parks and amenities

Inner Suburbia:
- Large low/mixed-density residential developments
- Streets and road grids
- Small-scale commerce
- Neighbourhood-scale public spaces

The Urbanised Countryside (Zwischenstadt):
- Low-density/scattered residential developments
- Motorways and regional-scale road grids
- Industries
- Small towns/satellite towns/rural villages
- Farming lands
- Landfills and brownfields
- Big architectonic artifacts (airports, power plants, others)
- Large natural reserves and regional parks
- Conurbation zones
- Satellite towns

The interstitial spaces

Any form of growth restriction

Any form of transport infrastructure

The open countryside:
- Extensive agricultural functions
- Protected natural reserves

Figure 2.2 The three scopes in which the interstitial spaces can manifest.

Source: Author.

restricts the access to a site crossed by electric lines. These interstices used to be 'surrounded' by built-up space and thus, they are in tension with adjacent land uses such as residential (Figure 2.3). A low-density area surrounded by highly densified surroundings can also be an interstice as it appears as a gap in the urban fabric in terms of density rate.

These inner interstices are recognised as spaces that are socially contested and reclaimed as part of the suburban fabric. They are identified as places with unexplored ecological contents and social benefits. The literature on ecosystem services (Douglas, 2008; Green et al., 2016), natural capital (Wang et al., 2012) and urban political ecology (Heynen et al., 2006) identify these interstices as potential 'green infrastructure' and elements that feed narratives of suburban retrofitting (Dunham-Jones & Williamson, 2009; Rice, 2010) and sustainability (Ma & Haarhoff, 2015). The environmental values of inner suburban interstices suggest a more sustainable development if codified in planning and included in debates around urban and human ecologies, as these interstices are not 'more or less "natural" than any other type of modern landscape' (Gandy, 2018, p. 78). This idea resonates with Naveh's notion of 'holistic landscape ecology', who argues for a more humanistic approach to the environmental conception of the spaces that can be integrated into the built environment, while involving both natural and societal processes alike (Naveh, 2000).

Figure 2.3 An infrastructural interstitial space in Paris' suburbs, France.

Source: Author, 2016.

Most green spaces in the inner city and inner suburban locations are protected and seldom changed (private gardens, private community squares, or small parks nearby). Yet, brownfield interstitial spaces in the same inner-city locations may be subject to planning 'blight' despite their latent economic value. Here any growth machine logic is one in which local and national state expenditures rather than purely private interests may come to the fore (Phelps & Wood, 2011). The 'rent gap' associated with derelict or unused land and properties (Smith, 1982) may not close without planning intervention (to clean up or provide access to sites) and even then may necessitate an urban politics and planning permissive of informal, temporary, or interim uses where decline is prolonged (Dubeaux & Sabot, 2018).

2.5.2 Contiguous expansion

While processes of extended suburbanisation continuous, new interstitial spaces emerge as part of. While originally located outside the city, they become interstitial when the sprawling growth spans them. Many of these spaces are indeed 'planned' by public agencies as part of the future areas of expansion; others are previously implemented as part of regional processes of economic development, or included in land-banking exercises by private agencies that keep them undeveloped for future development. Other areas are part of development plans labelled 'areas for expansion'. There is no specific pattern of location for these extension areas, but they used to be near suburban surroundings or linked to transport infrastructure or services (Camagni et al., 2002). These are the cases of originally outer industrial sites that are engulfed by sprawling growth over the years, or a natural reserve that becomes included into the development plan of a suburban district, or geographic restrictions – such as hills, valley, or a river basin – that are then transformed into metropolitan parks. As such, these interstitial spaces present a multi-level issue for governance, implying a need for coordination across different policy sectors and municipalities, most commonly in a monocentric metropolitan context. These interstitial spaces are usually the outcome of centralised decisions and administration, but with impacts at local levels. Yet, in this way, Talen (2010) argues that zoning contributes to sprawl, since open spaces are randomly distributed over the area of urban expansion due to the importance of previous infrastructures: 'There are many examples of this phenomenon, such as residential zones adjacent to eight-lane freeways, and public amenities surrounded by low-density, single-family zoning. In most cases, a more appropriate spatial pattern would put open space or more resilient uses adjacent to freeways, and higher-intensity land uses adjacent to public amenities' (Talen, 2010, p. 179).

In some cases, policymakers protect well-located lands for the future provision of services and the achievement of political goals. These are the cases of undeveloped areas that – because of their location and land capacity – appear as suitable for social-housing developments. In a similar vein, private sectors also examine these undeveloped lands as opportunities for further development, promoting land-use changes or the acquisition of these lands 'in-advance'. In these cases, these lands become privatised and often left undeveloped, a way of catching value over time (windfall gain). In other cases, these interstitial spaces are used as platforms of urban design and development projects that increase their land value to then being used as the subject of bigger mortgages or bank loans for parallel investments. These interstices attract the interest of both public and private sectors alike – not only because of their present conditions but also their future changes and opportunities – including their use as financial instruments (Morandé et al., 2010). These spaces are elements of dispute between local and central agencies, environmental groups, social-based organisations, and private developers, which entails their complexity in terms of governance.

As part of this exercise of inclusion of interstitial spaces into the urban development, the contiguous expansion is subject of speculations of different sorts, and thus, subject of detailed scrutiny regarding the infrastructural qualities and ongoing development plans. Prime network spaces – such as conurbation zones – are often considered for implementing major transportation and communication hubs, including airports and distribution centres, and also large-scale neighbourhoods. These areas are supplied by roads and become ready for development, although they could originally emerge as restrictions zones. Examples of regional conurbations in Europe – such as the London-Cambridge and Stockholm-Uppsala corridors – describe 'bicentric urban systems' where relational linkages ultimately consolidate 'corridors cities' that again confirm the governance challenges of this regional in-between space of contiguous expansion (Batten, 1995).

2.5.3 *The urbanised countryside*

The 'urbanized countryside' (Sieverts, 2003) describes the peri-urban scope of interstitial spaces. However, the 'peri-urban' is a limited term for this case as it mainly refers to the edge of cities and the dynamics of the urban-rural transect. Considering that urban sprawl is a somehow un-bounded phenomenon – and that interstitial spaces can reach geographic extensions beyond the city itself – the term 'urbanised countryside' appears as more embracing of the elastic character of the sprawling growth and the interstices. The 'urbanised countryside' includes the peri-urban, not only includes the open countryside but also embraces the urban condition present beyond the edges of cities. As Sieverts indicates, questions around the boundaries of the urbanised countryside are still open. The urbanised countryside can be constrained to modes of land management and governance linked to the city, but also regional governance in which the peri-urban is extended far beyond. Discussions around the metropolitan scope of city-regions imply national levels of governance (Calzada, 2015; Scott, 2019), although the economic geography of the regional space describes planetary dynamics linked to global competitiveness (Kunzmann, 2004), mobility, trans-national transference of policy, knowledge and services (Arbabi et al., 2019), and environmental impacts associated to global warming and climate change (Sun et al., 2020).

The urbanised countryside is a space where an intersection of actors, modes of governance, economic and environmental processes, and differing urban-rural development interests take place in the production of the built-up lands and interstitial spaces. Hebbert (1986), for example, emphasises how the transitional zone surrounding cities 'displays a mixture of uses and building types interspersed with agricultural and vacant land' (p. 141), while Buxton and Choy (2007) noticed that the peri-urban is a mixed of varied land-uses and boundaries where strategic governance becomes critical for the sustainable management of the economic and environmental impacts of agriculture and urban development (Buxton & Choy, 2007).

According to Tacoli (1998), the 'urbanised countryside' is a problematic zone of transition because urban and rural activities amalgamate in the same geographical space, which finally defines a hybrid zone where agricultural and non-agricultural activities are spatially and functionally (de)integrated (Tacoli, 1998). This is an ambiguous space where urban sprawl can manifest as traditional homogeneous suburbanisation while interspersed with agricultural functions and other heterogeneous infrastructures that extend beyond what we associate to the city. A hybrid landscape defined by 'the merging of nature and culture extended to the rural within the urban, the agricultural within the industrial, so forth. Such mosaics of activities and land use patchworks are frequent in most peri-urban landscapes and make them different from cityscapes, suburbs, and satellite towns. They are spaces in-between suburbs and villages beyond metropolitan regions' (Leontidou & Couch, 2007, p. 244). The urbanised countryside is a place where a diversity of forces and interests – including those operating upon interstitial spaces – connect to development projects of regional and national interest that influences transitions in rural land-uses.

The morphology of the urbanised countryside is partly a reflection of 'leapfrogging' development; non-contiguous urbanisation made up by the presence of those in-between open tracts such as farmlands, forest lands, and other non-urbanised areas. The fringe belts and expansion areas are always a sort of 'pending' space under the logic speculation surrounding land development – though this is rarely a feature emphasised explicitly. Their condition centres on their accessibility or location relative to infrastructure and existing development. An example of this is the Randstad in the Netherlands (Bruinsma et al., 1993), or other in-between spaces of city-regions that describe a chain of interconnected villages, small towns, rural agglomerations, larger cities, and interstitial spaces such as the conurbation that interconnect Liverpool, Manchester, Huddersfield, and Leeds in the UK. The morphology of the urbanised countryside in particular reflects the enduring obstacles to, different cycles of, and innovation in, building activity over time (Whitehand, 2001), but also its regional salience that reaches larger geographic spaces of transitions between urbanised regions.

The urbanised countryside also talks about the planning restrictions around the preservation of rurality and the open countryside that creates planned interstitial spaces. These are the cases of green belts and other restrictions to sprawling growth. Aside from differing positions – as many argue that growth restrictions influence further leapfrogged development – green belts are still widely accepted regarding the preservation of valuable ecosystem services (Hong & Guo, 2017; Itkonen et al., 2015; Zepp, 2018), provision of green infrastructure (Albert & Von Haaren, 2017; Amati & Taylor, 2010), protection of open space for communities (Siedentop et al., 2016), and positive influence relative to home prices of nearby communities (Asabere & Huffman, 2009; Herath et al., 2015; Jun & Kim, 2017). This is important considering that productivist rural activities in the urbanised countryside have transitioned from monofunctional to multifunctional (post-productivist), as various policies and pressures have affected the continuous operation of farms in the urban fringe. This has also affected the preservation of natural aesthetics, attraction of new population, indigenous and cultural reserves, forest exploitation, mining (such as coal and gold), and tourism (Choy & Buxton, 2013; Roche & Argent, 2015). These restrictions evince the dual planned and unplanned nature of interstices of the fringe belt space of city-regions.

In the case of Australia, for instance, 'multifunctionality highlighted the fact that the forces of production, consumption and protection could, and in fact did, coexist, the relative strength of each force dictating the overall trajectory of that rural space' (Roche & Argent, 2015, p. 633). The historical productive logic of the countryside may consider opportunities supported by its new stage of economic and environmental links with the city's growth. So, the 'urbanized countryside' must shift the way of traditional functions based on exclusive farming productivity to activities that decrease natural processes (areas with controlled water levels and insecticides) and ecosystem disruptions, surpluses of manure, cattle diseases, etc. It implies that the agricultural sector in urban fringes ought to focus more on food quality, environmental processes, and more sustainable ecosystems rather than traditional extensive cultivation (van Leeuwen & Nijkamp, 2006). The transitional character of multifunctional rural fringe areas is a global response from regulatory zoning to achieve the harmonisation of urban growth and protection of the countryside. The 'Finger Plan' of Copenhagen, for instance, since its origins was oriented to safeguard important fragments of the countryside while promoting urban growth (Caspersen et al., 2006; Gravsholt Busk et al., 2006). This balance is morphologically defined by the creation of densified zones along transport corridors ('fingers') interspersed with interstitial spaces in the form of 'wedges' that penetrate the city up to its main core. These protected open tracts are composed of natural landscapes, ecological corridors, farmlands, and temporary low-density activities related to leisure and recreation (Vejre et al., 2007). Another example is the Oregon's comprehensive 'Urban Growth Boundary' (UGB) approach. The policy has been discussed extensively regarding impacts on housing prices (Lang & Hornburg, 1997; Xu & Zhu, 2018) but is recognised as the outcome of the rise of specific forms of environmental governance in Oregon, and subsequently, Portland. Since the 1950s, increasing environmental awareness finally 'forced policymakers at the federal, state, and local levels to enact measures that attempted to regulate (urban) growth and assure an anxious public that the health of themselves and the environment was a top priority' (Huber & Currie, 2007, p. 712). By the 1970s, political authorities echoed the claims of organised campaigners – including coalitions of farmers and environmentalists – to set multiple goals governing urbanisation and the preservation of productive lands, championing 'the aggressive protection of the state's natural beauty and easy access to natural resources from a rising tide of urban sprawl' (ibid, pp. 713–714), finally countering the threat of unnatural and unchecked urban development.

2.6 Conclusions

Urban sprawl is intrinsically embedded within contemporary reflections around the urban condition of our contemporary society and its planetary scope. As such – and while referring to the scope of cities – traditional sprawling suburbanisation and advanced post-suburban growth observed in heavily urbanised regions acknowledge the interstitiality as part of their urban character. Indeed, urban sprawl is affected by the presence and dynamics of interstitial spaces while defining a sprawl index or the alternative planning pathways to become built-up space. This all-embracing context is better clarified by Sieverts' *Zwischenstadt*: a disclosure of the intrinsic nature of urban sprawl. Nevertheless, the interstitial spaces that are part of open further questions about the boundaries of the regional hinterland of the *Zwischenstadt*, and the underlying assumption that urban sprawl is nevertheless an 'urban' issue limited to the geography cities.

Through the different descriptions provided by Sieverts around the *Zwischenstadt*, it has been emphasised that urban sprawl can hardly be considered without reference to the interstitial spaces that lie between developments. These interstices describe a varied geography in which multifunctionality and their planned and unplanned nature evinces the rationales that operate upon both the built-up and the interstitial space. The *Zwischenstadt* has also drawn attention to the regional and architectonic scales at which interstitial spaces are apparent, including the inner suburbia, the contiguous expansion, and the urbanised countryside.

Interstitial spaces can be read in terms of historical processes, which render them pending. Such potentials are most obvious with regard to the economic interests at play and that most forcefully suggest the inevitability of their transition to developed land. Nevertheless, there can also be important environmental, social, and political meanings and potentials invested in such spaces. Some of these have come to the fore in the remaking of urban nature and the politics of collective consumption. While highlighting the processes of sprawling growth as those that also define the fortunes of interstitial spaces.

The interstitial spaces evince the conceptual constraints for understanding urban sprawl as it is usually filtered through extant urban theories focused on intensive processes of 'the production of the space' as synonym of *built-up*. Yet urban sprawl is corseted by existing (and somehow obsolete) analytical containers of the 'urban' that minimise its comprehension from rural and regional theories. This undermines the inter-institutional relevance of sprawling processes that are administratively 'urban' but retain their rural (regional) structure and characteristics; characteristics that are unveiled by the dynamic of the interstitial spaces.

References

Abubakar, I. R., & Doan, P. L. (2017). Building new capital cities in Africa: Lessons for new satellite towns in developing countries. *African Studies, 76*(4), 546–565.

Albert, C., & Von Haaren, C. (2017). Implications of applying the green infrastructure concept in landscape planning for ecosystem services in peri-urban areas: An expert survey and case study. *Planning Practice & Research, 32*(3), 227–242.

Altieri, L., Cocchi, D., Pezzi, G., Scott, E. M., & Ventrucci, M. (2014). Urban sprawl scatterplots for urban morphological zones data. *Ecological Indicators, 36*, 315–323.

Amati, M., & Taylor, L. (2010). From green belts to green infrastructure. *Planning Practice & Research, 25*(2), 143–155.

Arbabi, H., Mayfield, M., & McCann, P. (2019). On the development logic of city-regions: Inter-versus intra-city mobility in England and Wales. *Spatial Economic Analysis, 14*(3), 301–320.

Asabere, P. K., & Huffman, F. E. (2009). The relative impacts of trails and greenbelts on home price. *The Journal of Real Estate Finance and Economics, 38*(4), 408–419.

Bagheri, B., & Tousi, S. N. (2018). An explanation of urban sprawl phenomenon in Shiraz Metropolitan Area (SMA). *Cities, 73*, 71–90.

Balducci, A. (2017). New tasks and new forms for comprehensive planning in Italy. In L. Albrechts, J. Alden, & Artur Da Rosa Pires (Eds.), *The changing institutional landscape of planning* (pp. 170–192). Routledge.

Barrington, C., & Millard, A. (2015). A century of sprawl in the United States. *Proceedings of the National Academy of Sciences, 112*(27), 8244–8249.

Batten, D. (1995). Network cities: Creative urban agglomerations for the 21st century. *Urban Studies, 32*(2), 313–327.

Beck, U., Bonss, W., & Lau, C. (2003). The theory of reflexive modernization: Problematic, hypotheses and research programme. *Theory, Culture & Society, 20*(2), 1–33.

Boland, P., Bronte, J., & Muir, J. (2017). On the waterfront: Neoliberal urbanism and the politics of public benefit. *Cities, 61*, 117–127.

Bontje, M. (2004). From suburbia to post-suburbia in the Netherlands: Potentials and threats for sustainable regional development. *Journal of Housing and the Built Environment, 19*(1), 25–47.

Borsdorf, A., Hidalgo, R., & Sanchez, R. (2007). A new model of urban development in Latin America: The gated communities and fenced cities in the metropolitan areas of Santiago de Chile and Valparaíso. *Cities, 24*(5), 365–378.

Bradbury, M., & Bogunovich, D. (2013), 'Resilient sprawl: An alternative Auckland plan' (paper delivered to the Governing City Futures Conference, Sydney, 16–17 August).

Braimoh, A. K., & Onishi, T. (2007). Spatial determinants of urban land use change in Lagos, *Nigeria. Land Use Policy, 24*(2), 502–515.

Branch, M. (2018). *Comprehensive city planning: Introduction & explanation.* Routledge.

Bruegmann, R. (2005). *Sprawl: A compact history.* University of Chicago Press.

Bruinsma, F., Pepping, G., & Rietveld, P. (1993). Infrastructure and urban development; The case of the Amsterdam orbital motorway. *Serie research memoranda.* Faculteit der Economische Wetenschappen en Econometrie. Vrije Universiteit Amsterdam.

Burger, M., & Meijers, E. (2012). Form follows function? Linking morphological and functional polycentricity. *Urban Studies*, *49*(5), 1127–1149.

Buxton, M., & Choy, D. L. (2007). Change in peri-urban Australia: implications for land use policies. *The state of Australian cities conference (SOAC 2017)* (pp. 291–302).

Calthorpe, P., & Fulton, W. (2001). *The regional city*. Island Press.

Calzada, I. (2015). Benchmarking future city-regions beyond nation-states. *Regional Studies, Regional Science*, *2*(1), 351–362.

Camagni, R., Gibelli, M. C., & Rigamonti, P. (2002). Urban mobility and urban form: The social and environmental costs of different patterns of urban expansion. *Ecological Economics*, *40*(2), 199–216.

Caspersen, O. H., Konijnendijk, C. C., & Olafsson, A. S. (2006). Green space planning and land use: An assessment of urban regional and green structure planning in Greater Copenhagen. *Geografisk Tidsskrift-Danish Journal of Geography*, *106*(2), 7–20.

Choy, D. L., & Buxton, M. (2013). Farming the city fringe: Dilemmas for peri-urban planning. In Q. Farmar-Bowers, V. Higgins, & J. Millar (Eds.), *Food security in Australia*. Springer.

Couch, C., & Karecha, J. (2006). Controlling urban sprawl: Some experiences from Liverpool. *Cities*, *23*(5), 353–363.

Curran-Cournane, F., Cain, T., Greenhalgh, S., & Samarsinghe, O. (2016). Attitudes of a farming community towards urban growth and rural fragmentation – An Auckland case study. *Land Use Policy*, *58*, 241–250.

Dockerill, B., & Sturzaker, J. (2019). Green belts and urban containment: The Merseyside experience. *Planning Perspectives*, *35*(2), 583–608.

Douglas, I. (2008). Environmental change in peri-urban areas and human and ecosystem health. *Geography Compass*, *2*(4), 1095–1137.

Dubeaux, S., & Sabot, E. C. (2018). Maximizing the potential of vacant spaces within shrinking cities, a German approach. *Cities*, *75*, 6–11.

Dunham-Jones, E., & Williamson, J. (2009). *Retrofitting suburbia: Urban design solutions for redesigning suburbs*. Wiley.

Ewing, R., Pendall, R., & Chen, D. D. (2002). *Measuring sprawl and its impact*. Smart Growth America.

Fishman, R. (1987). *Bourgeois utopias: The rise and fall of suburbia*. Basic Books.

Gallent, N., & Shaw, D. (2007). Spatial planning, area action plans and the rural-urban fringe. *Journal of Environmental Planning and management*, *50*(5), 617–638.

Galster, G., Hanson, R., Ratcliffe, M., Wolman, H., Coleman, S., & Freihage, J. (2001). Wrestling sprawl to the ground: Defining and measuring an elusive concept. *Housing Policy Debate*, *12*(4), 681–717.

Gandy, M. (2018). Urban nature and the ecological imaginary. In C. Lindner, & M. Meissner (Eds.), *The Routledge companion to urban imaginaries* (pp. 78–89). Routledge.

Garreau, J. (1991). *Edge city: Life on the new frontier* (1st ed.). Doubleday.

Gillham, O. (2002). *The limitless city: A primer on the urban sprawl debate*. Island Press.

Gravsholt Busk, A., Kristensen, S., Praestholm, S., Reenberg, A., & Primdahl, J. (2006). Land system changes in the context of urbanization: Examples from the peri-urban area of Greater Copenhagen. *Danish Journal of Geography*, *106*(2), 21–34.

Green, O., Garmestani, A., Albro, S., Ban, N., Berland, A., Burkman, C., & Shuster, W. D. (2016). Adaptive governance to promote ecosystem services in urban green spaces. *Urban Ecosystems, 19*(1), 77–93.

Hebbert, M. (1986). Urban sprawl and urban planning in Japan. *The Town Planning Review, 57*(2), 141–158.

Heinrichs, D., Lukas, M., & Nuissl, H. (2011). Privatisation of the fringes – A Latin American version of post-suburbia? The case of Santiago de Chile. In N. A. Phelps, & F. Wu (Eds.), *International perspectives on suburbanization: A post-suburban world?* (pp. 101–121). Palgrave Macmillan.

Herath, S., Choumert, J., & Maier, G. (2015). The value of the greenbelt in Vienna: A spatial hedonic analysis. *The Annals of Regional Science, 54*(2), 349–374.

Hess, G., Daley, S. S., Dennison, B. K., Lubkin, S. R., McGuinn, R. P., Morin, V. Z., & Wrege, B. M. (2001). Just what is sprawl, anyway. *Carolina Planning, 26*(2), 11–26.

Heynen, N., Kaika, M., & Swyngedouw, E. (2006). Urban political ecology. In N. Heynen, M. Kaika, and E. Swyngedouw (Eds.), *The nature of cities: Urban political ecology and the politics of urban metabolism* (pp. 1–20). Routledge.

Hong, W., & Guo, R. (2017). Indicators for quantitative evaluation of the social services function of urban greenbelt systems: A case study of Shenzhen, China. *Ecological Indicators, 75*, 259–267.

Horn, A., & Van Eeden, A. (2018). Measuring sprawl in the Western Cape Province, South Africa: An urban sprawl index for comparative purposes. *Regional Science Policy & Practice, 10*(1), 15–23.

Huang, X., Li, Y., & Hay, I. (2016). Polycentric city-regions in the state-scalar politics of land development: The case of China. *Land Use Policy, 59*, 168–175.

Huber, M. T., & Currie, T. M. (2007). The urbanization of an idea: Imagining nature through urban growth boundary policy in Portland, Oregon. *Urban Geography, 28*(8), 705–731.

Insch, A. (2018). Auckland, New Zealand's super city. *Cities, 100*(80), 38–44.

Insch, A., & Walters, T. (2018). Challenging assumptions about residents' engagement with place branding. *Place Branding and Public Diplomacy, 14*(3), 152–162.

Itkonen, P., Viinikka, A., Heikinheimo, V., & Kopperoinen, L. (2015). ES GreenBelt–A preliminary study on spatial data and analysis methods for assessing the ecosystem services and connectivity of the protected areas network of the green belt of Fennoscandia. Reports of the Ministry of the Environment. Ministry of the Environment. Department of the Natural Environment.

Jaret, C., Ghadge, R., Williams Reid, L., & Adelman, R. (2009). The measurement of suburban sprawl: An evaluation. *City & Community, 8*(1), 65–84.

Johnson, M. P. (2001). Environmental impacts of urban sprawl: A survey of the literature and proposed research agenda. *Environment and Planning A, 33*(4), 717–735.

Jun, M. J., & Kim, H. J. (2017). Measuring the effect of greenbelt proximity on apartment rents in Seoul. *Cities, 62*, 10–22.

Kearns, R. A., & Lewis, N. (2019). City renaming as brand promotion: Exploring neoliberal projects and community resistance in New Zealand. *Urban Geography, 40*(6), 870–887.

Kim, G., Miller, P. A., & Nowak, D. J. (2018). Urban vacant land typology: A tool for managing urban vacant land. *Sustainable Cities and Society, 36*, 144–156.

Knox, P. (2018). Reflexive neoliberalism, urban design, and regeneration machines. In T. Haas, & H. Westlund (Eds.), *In the post-urban world. Emergent transformations of cities and regions in the innovative global economy* (pp. 82–96). Routledge.

Kunzmann, K. R. (2004). An agenda for creative governance in city regions. *disP – The Planning Review*, *40*(158), 5–10.

Lang, R. (2002). Does Portland's urban growth boundary raise house prices? Housing policy Debate, *13*(1), 1–5.

Lang, E. (2003). *Edgeless cities: Exploring the elusive metropolis*. Brookings Institution Press.

Lang, R., & Hornburg, S. P. (1997). Planning Portland style: Pitfalls and possibilities. *Housing Policy Debate*, *8*(1), 1–10.

Leontidou, L., & Couch, C. (2007). Urban sprawl and hybrid cityscapes in Europe: Comparisons, theory construction and conclusions. In C. Couch, G. Petschel-Held, & L. Leontidou (Eds.), *Urban sprawl in Europe: Landscapes, land-use change and policy* (pp. 242–267). Blackwell Publishing.

Li, G., & Li, F. (2019). Urban sprawl in China: Differences and socioeconomic drivers. *Science of the Total Environment*, *673*, 367–377.

Li, L. Y., Qi, Z. X., & Xian, S. (2020). Decoding spatiotemporal patterns of urban land sprawl in Zhuhai, China. *Applied Ecology and Environmental Research*, *18*(1), 913–927.

Ma, J., & Haarhoff, E. (2015). The GIS-based research of measurement on accessibility of green infrastructure-a case study in Auckland. In *MIT CUPUM Conference Proceeding, The 14th International Conference on Computers in Urban Planning and Urban Management, July 7–10*. MIT Press.

MacLachlan, A., Biggs, E., Roberts, G., & Boruff, B. (2017). Urban growth dynamics in Perth, Western Australia: Using applied remote sensing for sustainable future planning. *Land*, *6*(1), 9.

Masoumi, H. E., Hosseini, M., & Gouda, A. A. (2018). Drivers of urban sprawl in two large middle-Eastern countries: Literature on Iran and Egypt. *Human Geographies*, *12*(1), 55–79.

Morandé, F., Petermann, A., & Vargas, M. (2010). Determinants of urban vacant land. *The Journal of Real Estate Finance and Economics*, *40*(2), 188–202.

Morrison, N. (2010). A green belt under pressure: The case of Cambridge, England. *Planning Practice & Research*, *25*(2), 157–181.

Naveh, Z. (2000). What is holistic landscape ecology? A conceptual introduction. *Landscape and Urban Planning*, *50*(1–3), 7–26.

Nelson, A. (1999). Comparing States with and without growth management analysis based on indicators with policy implications. *Land Use Policy*, *16*, 121–127.

Odum, E., & Barret, G. (2006) Fundamentals of ecology. Cengage Learning.

Oueslati, W., Alvanides, S., & Garrod, G. (2015). Determinants of urban sprawl in European cities. *Urban Studies*, *52*(9), 1594–1614.

Phelps, N. (2012). *An anatomy of sprawl. Planning and politics in Britain*. Routledge.

Phelps, N. (2018). In what sense a post-suburban era? In B. Hanlon, & T. Vicino (Eds.), *The Routledge companion to the suburbs* (pp. 39–47). Routledge.

Phelps, N. A. (2015). *Sequel to suburbia: Glimpses of America's post-suburban future*. MIT Press.

Phelps, N. A., & Silva, C. (2017). Mind the gaps! A research agenda for urban interstices. *Urban Studies*, *55*(6), 1203–1222.

Phelps, N., & Wood, A. (2011). The new post-suburban politics? *Urban Studies*, *48*(12), 2591–610.

Phelps, N., Wood, A., & Valler, D. (2010). A postsuburban world? An outline of a research agenda. *Environment and Planning*, *42*(01), 366–383.

Pirotte, A., & Madre, J. L. (2011). Determinants of urban sprawl in France: An analysis using a hierarchical Bayes approach on panel data. *Urban Studies, 48*(13), 2865–2886.

Ponzini, D. (2014). The values of starchitecture: Commodification of architectural design in contemporary cities. *Organizational Aesthetics, 3*(1), 10–18.

Qian, H., & Wong, C. (2012). Master planning under urban-rural integration: The case of Nanjing, China. *Urban Policy and Research, 30*(4), 403–421.

Rauws, W. S., & de Roo, G. (2011). Exploring transitions in the peri-urban area. *Planning Theory & Practice, 12*(2), 269–284.

Rice, L. (2010). Retrofitting suburbia: Is the compact city feasible? *Proceedings of the Institution of Civil Engineers-Urban Design and Planning, 163*(4), 193–204.

Roche, M., & Argent, N. (2015). The fall and rise of agricultural productivism? An antipodean viewpoint. *Progress in Human Geography, 39*(5), 621–635.

Røe, P. G., & Saglie, I. L. (2011). Minicities in suburbia – A model for urban sustainability? *FormAkademisk-forskningstidsskrift for design og designdidaktikk, 4*(2), 38–58.

Salet, W., & Woltjer, J. (2009). New concepts of strategic spatial planning dilemmas in the Dutch randstad region. *International Journal of Public Sector Management, 22*(3), 235–248.

Scott, A. J. (2019). City-regions reconsidered. *Environment and Planning A: Economy and Space, 51*(3), 554–580.

Siedentop, S., Fina, S., & Krehl, A. (2016). Greenbelts in Germany's regional plans – an effective growth management policy? *Landscape and Urban Planning, 145*, 71–82.

Sieverts, T. (2003). *Cities without cities. An interpretation of the Zwischenstadt.* Spon Press.

Sieverts, T. (2011). The in-between city as an image of society: From the impossible order towards a possible disorder in the urban landscape. In D. Young, P. Wood, & R. Keil (Eds.), *In-between infrastructure: Urban connectivity in an age of vulnerability* (pp. 19–27). Praxis (e)Press

Silva, C. (2019). Auckland's urban sprawl, policy ambiguities and the peri-urbanisation to Pukekohe. *Urban Science, 3*(1), 1–20.

Silva, C., & Vergara-Perucich, F. (2021). Determinants of urban sprawl in Latin-America: Evidence from Santiago de Chile. *SN Social Sciences, 1*(8), 1–35.

Smith, N. (1982). Gentrification and uneven development. Economic Geography, *58*(2), 139–155.

Soule, D. (2006). Defining and managing sprawl. In D. C. Soule (Ed.), *Urban sprawl: A comprehensive reference guide.* Greenwood Publishing Group, Inc.

Ståhle, A., & Marcus, L. (2008). Compact sprawl experiments. Four strategic densification scenarios for two modernist suburbs in Stockholm. In A. Stahle (Ed.), Compact sprawl: Exploring public open space and contradictions in urban density. Published Doctoral Dissertation, KTH Architecture and the Built Environment, School of Architecture, Royal Institute of Technology.

Sun, B., Han, S., & Li, W. (2020). Effects of the polycentric spatial structures of Chinese city regions on CO_2 concentrations. *Transportation Research Part D: Transport and Environment, 82*, 102333.

Tacoli, C. (1998). Rural-urban interactions: A guide to the literature. *Environment and Urbanization, 10*(1), 147–166.

Talen, E. (2010). Zoning for and against sprawl: The case for form-based codes. *Journal of Urban Design, 18*(2), 175–200.

Tapia, R. (2013). Evolution of the spatial pattern of social housing in Gran Santiago, Chile. 1980–2010. *Revista Geográfica Venezolana*, *55*(2), 255–274.

Tarazona Vento, A. (2017). Mega-project meltdown: Post-politics, neoliberal urban regeneration and Valencia's fiscal crisis. *Urban Studies*, *54*(1), 68–84.

Trubka, R., Newman, P., & Bilsborough, D. (2010). The costs of urban sprawl – infrastructure and transportation. *Environment Design Guide*, *83*(1), 1–6.

van Leeuwen, E., & Nijkamp, P. (2006). The urban-rural nexus; A study on extended urbanization and the hinterland. *Studies in Regional Science*, *36*(2), 283–303.

Vejre, H., Primdahl, J., & Brandt, J. (2007). The Copenhagen Finger Plan. Keeping a green space structure by a simple planning metaphor. Europe's living landscapes. Essays on exploring our identity in the countryside. *Landscape Europe*/KNNV.

Vergara, F., & Boano, C. (2020). Exploring the contradiction in the ethos of urban practitioners under neoliberalism: A case study of housing production in Chile. *Journal of Planning Education and Research*. https://doi.org/10.1177/0739456X20971684

Wang, Z., Nassauer, J. I., Marans, R. W., & Brown, D. (2012). Different types of open spaces and their importance to exurban homeowners. *Society & Natural Resources*, *25*(4), 368–383.

Weilenmann, B., Seidl, I., & Schulz, T. (2017). The socio-economic determinants of urban sprawl between 1980 and 2010 in Switzerland. *Landscape and Urban Planning*, *157*, 468–482.

Whitehand, J. (2001). British urban morphology: The Conzenian tradition. *Urban Morphology*, *5*, 103–109.

Wolman, H., Galster, G., Hanson, R., Ratcliffe, M., Furdell, K., & Sarzynski, A. (2005). The fundamental challenge in measuring sprawl: Which land should be considered? *The Professional Geographer*, *57*(1), 94–105.

Xu, J., & Zhu, Y. (2018). The impact of Portland's urban growth boundary. *2nd International Conference on Education Technology and Social Science* (ETSS 2018) (pp. 303–309).

You, H., & Yang, X. (2017). Urban expansion in 30 megacities of China: Categorizing the driving force profiles to inform the urbanization policy. *Land Use Policy*, *68*, 531–551.

Zepp, H. (2018). Regional green belts in the Ruhr region a planning concept revisited in view of ecosystem services. *Erdkunde*, *72*(1), 1–22.

Zhang, R., Pu, L., & Zhu, M. (2013). Impacts of transportation arteries on land use patterns in urban-rural fringe: A comparative gradient analysis of Qixia district, Nanjing City, China. *Chinese Geographical Science*, *23*(3), 378–388.

3 The interstitial spaces of urban sprawl

> Instead of talking dismissively about urban sprawl, we could recognise that
> there is a fine-grained interpenetration of open space and built form and
> see the open space as the binding element, with its new creative potential.
>
> (Sieverts, 2003, p. 49)

3.1 Introduction

It has been clarified that the urbanisation process cannot be solely understood from the contents and dynamics of the *built-up*. Processes of urbanisation can hardly be considered without reference to the undeveloped areas
and open tracts that lie between developments, above all in contexts of
sprawling growth characterised by land fragmentation. There is a spectrum
of undeveloped areas and open tracts produced alongside the built-up space
named '*interstitial spaces*' that are varied in nature, scale, functions, and
spatial characteristics. The extant literature that refers to these interstices,
however, is fragmented and quite singular in its treatment in spatial, ecological, or other terms. There is also no clarity in how the interstitial spaces
are produced – as they apparently emerge as less controlled processes in
planning – or what is the type of geography that they suggest, or what could
be the conceptual scaffolding that operationalise their analysis.

In this chapter, the different terms that refer to interstitial spaces are
critically discussed to then place the term *interstitial space* as a more
embracing conceptual framework for the varied meanings and types of
non-urban geographies between (and beyond) cities and regions. The term
interstitial space provides a unifying agenda of research for the whole
spectrum of undeveloped areas and open tracts that compose the interstitial geography of cities and the space of relation between cities and other
types of agglomerations. The *interstitial geography* is also proposed as the
wider context in which the interstitial spaces manifest and where different
interstitial enclaves and agglomerations – *the interstitial hubs* – articulate
different infrastructures of network and varied functional and environmental processes. The conclusions highlight the importance of interstitial

DOI: 10.4324/9780429320019-3

spaces and their underlying connections to the logics of the production of the space while distinguishing the component of the interstitial geography, network spaces, and the interstitial hubs. All these elements configure the scope in which some of the most radical political processes that contribute to urbanisation occur and where some of the most profound environmental alterations take place. These characteristics of the interstitial spaces, as well as their planned and unplanned nature, are also introduced to be further discussed in the next chapter.

3.2 Critical antecedents from a fragmentary literature

Considering the varied nature of interstitial spaces and the different determinants that explain their emergence, the extant literature the touches on such interstices is also varied, fragmented, and quite singular in its treatment in architectural, ecological, social, and other terms. These terms are specific to certain types of interstitial spaces and hardly provide a unifying perspective for the whole system that composes the interstitial geography that lies between (and beyond) sprawling cities and regions. Because of their specificity and strong empirical focus, these attempts tend to be conceptually partial or limited; a specific term used to describe an open and abandoned green space, for instance, cannot explain a regional park or conurbation zone between two cities. However, all these terms share a common ground around the 'in-between' condition of these spaces, which makes them conceptually familial. In other cases, existing terms have an architectonic ground mainly useful to describe small-scale spaces. These varied terms are nevertheless important for the exercise of unveiling the varied nature, potentials, commonalities, and scope of the interstitial geography.

'Undeveloped areas', undeveloped spaces, undeveloped lands, 'undevelopable' lands, and other similar terms are used to describe empty sites often defined by physical handicaps or geographical constraints that affect the continuous urbanisation of cities (Theobald, 2001; Wolman et al., 2005). From a development perspective, this term refers to these physically empty sites as gaps. A farming site located at the peri-urban space, for instance, can be agriculturally productive, open to the public and tourism, used for educational activities, leisure, or massive encounters, or serve as an informal backyard for suburban urban residents, but nevertheless described as an 'undeveloped land' as it does not show any type of densification. Similarly, a constructed space – such as an abandoned industrial facility or a building – can also be described as 'undeveloped' as it does not provide any sign of activity or integrated functions to the city (Batty, 2016).

The term 'vacant land' arose in the early twentieth century to describe a variety of undeveloped spaces in cities including geographical handicaps, undeveloped private or public properties, empty sites aimed to support future social services (schools, religious, and others). These sites are usually associated with previous industrial functions, often reclaimed for regeneration or infilling policies (Foo et al., 2014; Ige & Atanda, 2013; Northam, 1971). Their (obsolete) industrial past depicts their condition as 'vacant'; brownfield sites that are derelict and polluted. Underused, abandoned, derelict, and although fenced, these vacant lands are often informally occupied and partially destroyed or demolished (Pagano & Bowman, 2000). According to Northam (1971), vacant lands can be classified into four types. First, 'remnant parcels' that include slopes, flood valleys, and other small plots useful for implementing new public spaces. Second, 'corporate reserves' owned by business corporations to support future expansions. Third, 'speculative areas' can be sold or used as financial commodities, and fourth, 'open tracts' owned by public (or semi-public) institutions as reserves for social facilities. As such, the notion of 'vacant land' is somehow melted with the idea of 'undeveloped land' and used to identify empty/undeveloped sites with the potential to be urbanised.

The notion of 'open space' carries a positive connotation associated with social, environmental, and economic benefits. Open spaces are key elements for highly densified areas (Yuan et al., 2015). They are also implemented to reduce impacts of natural disasters (Barkasi et al., 2012). Concerning economics, open spaces are positive assets for new developments if they are nearby transport infrastructure or services (Bruinsma et al., 1993; Clawson, 1962; Graham, 2000). Socially and spatially, open spaces provide specific features that make places unique, and when they have integrated functions, encourage contact between people and reduces perceptions of insecurity (Koohsari et al., 2015). A more negative connotation is proposed by Loukaitou-Sideris (1996), who describes 'open spaces' as fractures, discontinuities, and socio-spatial disruptions of the urban fabric. Open spaces can be spatial barriers, boundaries, and residues; leftover spaces that do not contribute to the urban quality. Examples of these 'open spaces' are the parking areas between suburban corporate towers, airports, shopping malls, and others. Also, 'open spaces' can be abandoned (unsafe) parks and squares, similar to the open tracts resulting from the construction of motorways, commercial strips or industrial complexes.

De Solá-Morales' popular 'terrain vague' [vague land] is a well-established concept in the architectural literature. The term describes spaces and buildings alike in a double condition: First, they are 'vague' in the sense of empty, without activities or functions; unproductive spaces, abandoned, obsoletes or in a stage of ruin, and second, they are imprecise, undefined, (a 'form of absence') without fixed limits or future destinations irrespective of their previous signs of occupation (de Solá-Morales, 2002). The author highlights the importance of these spaces for the collective memory and social identity, often exemplified by derelict and abandoned industrial facilities that evoke past heydays while reinforcing socially accepted values. Also, many suburban empty sites are used as examples of 'terrain vagues', sites that were integrated gears of the suburban space that are now abandoned. It is worth noticing that the 'terrain vague' refers to both empty and built-up spaces alike.

The term 'wildscape' is used to describe abandoned sites in cities that show different degrees of wildlife (flora and fauna). It includes both open spaces and built-up areas such as abandoned buildings, ruins, or unattended facilities where wildlife has taken place. Jorgensen and Keenan (2012) define the 'wildscapes' as spaces or architectonic facilities where the city's forces of control have not defined the space at all. These are places where children and spontaneous activities can flourish, and unexpected forms of occupation can be developed beyond those determined by formal plans. The 'wildscapes' are seen as positive elements of the urban fabric – despite their certain degree of marginality and abandonment – because of their social and environmental contributions. These spaces are necessary to balance advanced levels of urbanisation in regions where residents are excluded from the experience with nature, flora, and fauna. Some examples of 'wildscapes' are lakes, forest lands, ecological reservoirs, but also abandoned sites where nature has spontaneously grown (Aurora et al., 2009; Jorgensen & Keenan, 2012; Kitha & Lyth, 2011).

Urban 'wastelands' are abandoned, marginalised, and forgotten urban spaces characterised by spontaneous and exuberant vegetation with notable aesthetics and ecological benefits. Urban wastelands are important habitats for many insects, although poorly protected by planning regulations. They are often found in areas affected by industrial activity, after the demolition of industrial buildings and warehouses, or near abandoned infrastructure (Twerd & Banaszak-Cibicka, 2019). Referring to these spaces, Gandy (2013) developed the term 'marginalia' to describe wastelands in cities like London, Berlin, and Montreal. These spaces emerge from former industrial facilities, landfills, and abandoned lands. They are also described as spaces for scientific exploration due to their ecological properties. 'Wastelands' offer strong sensorial stimulation considering their particular aesthetics, random expressions of freedom, spontaneity, and some hints of history and novelty at the same time. According to Gandy (2013), the ambivalent condition of 'wastelands' appears as prolific scenarios to combine new cultural and scientific expressions of urban life and 'as practical challenges for the utilitarian impetus of capitalist urbanization' (Gandy, 2013, p. 1312).

The term 'non-urbanised areas' (NUAs) describes natural urban landscapes with a clear ecological role. NUAs include agricultural lands and any kind of green/blue infrastructure such as parks, squares, rivers, canals, forested roads, and others without signs of intervention. As such, NUAs support environmental services considering their rich biochemical composition. Their relevance also relates to socio-economic benefits as public space, contact with nature, and spontaneous activities. La Greca et al. (2011) indicate that 'NUAs are strategic both for pollution minimization and climate change adaptation. In land-use planning, they should be carefully considered since they can give a contribution to the enhancement of urban settlement quality and to the improvement of human health. Several environmental, social, economic and cultural benefits can derive from agriculture and green infrastructures in urban areas' (La Greca et al., 2011, p. 2201). NUAs are also utilitarian in the study of different forms of wildlife, considering their high levels of biodiversity, including endangered species found in urban forests. (Alvey, 2006). Planned NUAs – such as parks and squares – are criticised, however, for their artificialised aesthetics that alter natural properties and threaten their intrinsic character as ecological spaces. Gandy (2011) highlights the relevance of unplanned NUAs in Berlin as sources of biodiversity in contrast to those institutionally planned: 'the site is aesthetically and scientifically much more interesting than the closely managed municipal park to the north side of the street with its short turf and widely spaced trees. The urban meadow is an example of what the urban ecologist Ingo Kowarik terms *fourth nature* that has developed without any human design or interference to produce a new wilderness' (Gandy, 2011, p. 150). Aside from differing positions, the biodiversity defined by NUAs has positive impacts on the overall quality of life, education, and social culture (Savarda et al., 2000). There are several examples of green interstices managed by local residents, volunteers, community groups, and NGOs that support the provision of organic food, educational services, research activities, production of clean energy, and others (Saunders, 2011). NUAs reinforce the sense of safety (Stamps & Smith, 2002), provide spatial diversity, and contribute to the overall wellbeing by providing psycho-emotional and immaterial benefits in highly urbanised and homogeneous environments (Chiesura, 2004; Manzo, 2003).

The notion of 'non-place' (1995), coined by Marc Augé, has opened a wide range of interpretations regarding the tractability of the concept in empirical research on cities and urban studies. When used by architects, it refers to a spatial condition that cannot be properly identified as a 'place' but has nevertheless the potential to become a 'place'. In that sense, the 'non-place' is not the same as an 'anti-place' but an 'almost place' that, through architectonic interventions, can be integrated into the urban fabric. It is also used to describe transitional spaces; socio-spatial instances – somehow out of norms, plans and regulations – without clear definitions in terms of identity, character, or defined functions; spaces of mobility resulting from global processes of modernisation (Yuan et al., 2015). The notion of 'non-place', however, has more conceptual salience around the philosophical underpinning around the notions of 'space' and 'place'. The contribution to the understanding of the interstitial spaces from Augé's 'non-place' will be further developed in the next sections of this chapter.

The 'Drosscape' (Berger, 2006) refers to the array of spaces outside regulations and institutional norms. These spaces are described as opportunities for urban design and landscape interventions. They emerge as by-products of spatially fragmented suburbanisation; leftovers of the economic decline of some American cities. These are undervalued spaces due to their pollution, vacancy, and lack of economic value. These spaces emerge as spatial boundaries delinked from the functions of the city (Berger, 2006).

From an urban design perspective, Roger Trancik's notions of 'anti-space' and 'lost space' (1986) also attempt to describe interstitial spaces within the modern fabric of cities. Trancik argues that cities contain a number of 'anti-spaces' – shapeless, continuous, lacking perceivable form – resulting from modern planning practices subject to strict zoning regulations, land-use policies, and car-orientated designs. These anti-spaces are 'lost' in the sense of being leftovers, underutilised, unstructured, away from the flow of pedestrian activity, or abandoned; 'they are the residual areas between districts and loosely composed commercial strips that emerge without anyone realizing it' (Trancik, 1986, p. 3). 'Lost spaces' are abandoned waterfront, train yards, vacated military and industrial sites, and facilities that must be identified to 'then to fill them with a framework of buildings and interconnected open-space opportunities that will generate new investment' (ibid, p. 2).

Franck and Stevens' notion of 'loose space' (2006) refers to undetermined spaces that do not fit into well-established categories of public space with well-established functions. The authors argue that 'the variety of open spaces in cities includes those that are planned for certain assigned functions but that, both legally and physically, accommodate other activities as well; it also includes other kinds of spaces currently without assigned functions that accommodate unintended and unexpected activities' (Frank and Stevens, 2006, p. 4). The 'loose space' is defined as flexible and open to different purposes. Socially, 'loose spaces' are spaces of freedom of choice, lack of control, improvisation, and random encounters. These spaces contribute to the diversity and vitality of the city and are identified through the disjunction between the form of the space (and its intended functions) and the concrete activities that they host. This disjunction between 'form and function' has fed various debates in design and architectural circles, which, when extrapolated into the city-scale discussion, opens further reflections about the 'planning of the space' versus the 'planning of the use' of the space and its more spontaneous modes of social appropriation.

The notion of 'interplace' was used in the architectural field by Edward Casey in 1993, to describe spaces that connect interior and exterior spaces in buildings, such as balconies, porches, and others. These spaces can be open, semi-open, public, or semi-public as they support spatial and functional transitions between other more consolidated spaces to stay (in and out). These intermediate spaces are 'interplaces' that are 'between inner and outer as well as between front and back, right and left, and up and down, this interstitial structure does not importune the visitor to go either in or out' (Casey, 1993, p. 126). Casey's 'interplace' unveils both the three-dimensional (architectonic) character of the 'interplace' and its condition as a mediator (or bridge) between other spaces. Further developed by Phelps (2017), the notion of 'interplace' is then taken to the wider scale of regions and nations. Using an economic geography lens, the 'interplace' talks about the production of new spaces of accumulation situated in non-territorial forms of organisation. 'Interplaces' is a taxonomical attempt to describe 'the economy in between' developed between regions and nations. In these large-scale places, many economic activities are developed with impacts at national and global levels. Phelps discusses how *agglomerations, networks, enclaves,* and *arenas* define 'the coexistence of geographical formations that – as interplaces – lie somewhere in between' (Phelps, 2017, p. 8). Phelps argues that 'interplaces' describes new expressions of the production of the space where contemporary 'personal and virtual mobility has also brought forward an interest in non-territorial metaphors such as networks, rhizomes, flows, virtuality/simultaneity, as well as terms such "glocalization" that do indeed in their own way offer a picture of the interplace economy' (ibid, p. 8).

The 'interfragmentary space' was proposed by Vidal in 1999, who stated that 'the urban phenomenon is essentially a permanent tension between fragments' (Vidal, 2002, p. 150), and the space that lies between fragments would be the interfragmentary space. The 'fragments' are built-up units of different sorts (buildings, neighbourhoods, large architectonic artefacts) characterised by morphological, economic, and social dimensions, dimensions that would also be attached to the interfragmentary space. Although urban fragmentation has been associated with social segregation (Prévôt-Schapira & Cattaneo, 2008), Vidal argues that fragmentation is more a morphological process that supports the mobility of specific groups (a 'social fragment'). It means that the interfragmentary space could be a transitional space instead of a barrier that segregates the city: 'the field of relations where transitions between fragments are produced' (Vidal, 2002, p. 147); a space of 'reconciliation' that facilitates the recognition of the boundaries of fragments (ibid). Vidal also argues that the interfragmentary space does not have its own identity as it depends on the predominant dimensions of its surroundings. These dimensions can be structural, morphological, ecological, economic, or social, and thus, the interfragmentary space modifies its identity according to the transformations of its surroundings (Vidal, 2002). Vidal indicates that the interfragmentary space emerges from five types of interrelations between fragments: 'an overlap of fragments, a suture, a third overlapped element, an immersion of fragments and a space of networks' (Vidal, 2002, p. 152). The latter is used to describe spaces physically separated but connected by different infrastructures: 'it is suitable to consider "networks" as a fifth mode of inter-fragmentarity: fragments physically distanced but joined by lines, channels, pipes, elements in movements. The network allows the exchange of information, people, goods and circulation of commodities. It may get the form of communication channels or a strategic system of elements over the territory'

(Vidal, 2002, p. 156). Vidal's understanding of cities as composed of fragments supposes that 'fragments' can be spatially (and morphologically) identified. This exercise is possible to be done in more compact cities – or dispersed cities where fragments have clear boundaries nevertheless. However, this is not always the case – above all in contemporary sprawling landscapes – as boundaries between fragments tend to be diffuse, extended, and composed of eclectic infrastructures.

Although used in a wide range of disciplines, the terms 'interstitial space' and 'interstice' are rarely used in urban studies. When used, they are utilitarian to highlight spaces that emerge as 'gaps' within the built-up fabric of cities or to reinforce the economic importance of the wide geography between cities and regions. The terms are also used to describe the space in which alternative social interactions occur – often linked to informality and contestation of hegemonic politics of the city. Pflieger and Rozenblat (2010) use 'interstitial places' to describe spaces in cities as 'bridges' between formal institutional arrangements and alternative forms of social organisation. The authors argue that 'in a metropolis, these "bridges" can be (...) physical, like road or public transport systems (...) cultural or political, through "interstitial" places where different social groups converge' (p. 2727). Similarly, Brighenti (2013) emphasises the sociological aspects of interstices as gaps to be penetrated by the excluded in society. Schnell and Benjamini (2005) argue that 'interstices' are the spaces of integration and interaction between local and non-local groups. For Dovey (2012) and Shaw and Hudson (2009), interstices are spaces in which urban informality and creativity can flourish, though for Sousa-Matos (2009), these spaces are by-products to be re-incorporated into formal development processes. As part of their development and environmental potentials, Hugo and du Plessis (2020) suggest that planning can provide 'innovative solutions that lever existing unused and underutilized interstitial spaces within the urban fabric for climate change adaptation and mitigation purposes' (p. 591).

Insights from economic geography highlight the importance of the regional in-between space of capital accumulation by using the term 'interstitial space' or 'interstice' in different ways. Sayın et al. (2020), for instance, argue that within the context of globalisation there are 'interstitial cities' that count in the global circuits of capital accumulation. These 'interstitial cities' – such as Istanbul – are 'all located in an interstitial zone situated between Western Europe and Asia' (p. 12). Rickards et al. (2016) propose expanding the understanding of city-regions by focusing on those spaces beyond cities as the 'adoption of a "city region" lens obscures the interstitial spaces that lie between these designated metropolitan areas but that produce a large percentage of national wealth' (p. 1536). Similarly, Harrison and Heley (2015) argue that 'despite a select group of urban centres generating a disproportionate amount of global economic output, significant attention is being devoted to the impact of urban-economic processes on interstitial spaces lying between metropolitan areas' (p. 1113). The authors highlight the 'interstitial space' as rural areas that contribute to the prosperity of cities. Pemberton and Shaw (2012) also stress that 'spaces located between metropolitan areas 'are important spaces that cannot and should not be ignored (p. 446); interstitial spaces must be brought into closer conceptual focus. In a similar vein, Zhang and Grydehøj (2020) use the term 'urban interstices' while referring to 'the inter-urban boundary areas and interface zones that facilitate exchange between and within vast urban systems' (p. 1). They exemplified this through the case of Zhoushan Archipelago (Zhejiang Province, China) – and interface zone both between cities within the Yangtze River Delta urban agglomeration and between the Yangtze River Delta urban agglomeration and other megaregions; a space they refer to as an 'interstitial island' (p. 1).

As observed, these terms refer to very specific types of interstitial spaces. Their strong disciplinary focus somehow limits their explicative stand beyond architecture, urban design, economic geography, or planning for retrofitting or infilling. They refer to limited scales at which the interstices are manifested. It is difficult, for instance, to indicate that the regional space between two cities is simply a 'vacant land' or a 'terrain vague'; or that a farming area engulfed by the sprawling expansion is a 'lost space', or that a military facility is a 'wildscape'. When used in urban studies, the term 'interstitial space' appears as very generic and interchangeably used to refer to any kind of interstitial 'entity' between developments such as cities, places, spaces, regions, enclaves, agglomerations, and others; all entities that are different in nature, scale, morphology, relationality but mainly in their realities as 'spaces' or 'places'. It would be conceptually reductive to sustain that the economic focus on the interstitial spaces is enough to describe agglomerations that can indistinctively take the form of 'spaces' or 'places' alike. The previously discussed notions are nevertheless illustrative of the multifaceted character of the interstitial geography in both spatial and functional terms, but the term 'interstitial space' deserves closer inspection and a more stable position in urban studies. What is for sure, however, is that these terms suggest that it is no longer going to be acceptable to conceptualise the in-between space or the rural scope simply 'as an appendage hanging on to the coattails of the great modern metropolis' (Harrison & Heley, 2015, p. 1131).

3.3 The abstract grounds of the interstitial spaces

The interstitial spaces can be perceived as components of spatial and social segregation, disconnection, and isolation between different social groups (Foo et al., 2014); spaces that leave behind the classical notions of cities as continuous and ordered environments (Stavrides, 2007). This fragmentation – different from Vidal's notion of fragmentation (2002) and consequently linked to the loss of the unitary form of the built environment – has also been a matter of discussion in architecture, urban design, and planning circles where the presence of vacant lands, undeveloped areas, and other interstices are spatial interruptions to the urban fabric; places to be infilled, retrofitted, revamped, or reknitted for the sake of spatial consistency and functional continuity (Trancik, 1986).

Within these disciplinary fields – however – in 1995, the Dutch architect Rem Koolhaas proposed the term 'nothingness' to elucidate the spatial character of Berlin after WWII, a city configured by different (survival) built-up spaces interspersed with urban voids resulting from the destruction of the city. These voids – open scars where architecture or development of any kind was absent – amalgamated the fears of the past with the optimism of the city's reconstruction. Koolhaas indicated – however – that 'where there is nothing, everything is possible; where there is architecture, nothing (else) is possible' (Koolhaas, 1995, p. 199), alluding to the importance of the urban voids as part of the city, Koolhaas made an explicit claim against the institutional attempts to reknit the damaged urban fabric by filling the voids. Koolhaas contended that Berlin is now a divided, disperse, and decentralised city in which the urban voids express Berlin's new character of a dazed, fragmented, and destroyed land with no centre. He argued that the future of Berlin must operationalise the destruction (deconstruction) as mechanisms to (re)create the character of Berlin, the creation of new voids, and demolition of its dysfunctional parts. This would leave space for the wild forest to grow and create a place in which isolated (and somehow hidden/desolate) islands of infrastructures and ruins would be dispersed, floating in a large empty space of 'nothingness'; the new 'urban' condition for Berlin defined by unplanned and planned new interstitial spaces. Koolhaas' imaginaries on Berlin – and the inexorable implications of interstitial voids in the socio-political agenda – talk about the emergence of interstices ostensibly as a matter of intended design.

The view of the contemporary city as one of the fragments is not unique but the connotation of the term 'fragmentation' is. For many, 'urban fragmentation' amounts to forms of segregation and has negative social, economic, and political consequences (Monkkonen et al., 2018). However, Vidal (1999) highlights that the term 'fragmentation' can be explicative of the urban phenomenon if we take this apart from historical contingencies linked to socio-economic segregation. For Vidal, urban fragmentation is a process of 'addition or subtraction of different pieces which are part of a general urban trend' (Vidal, 1999, p. 158) yet also does not necessarily entail the physical coincidence of fragments and interdependency. A specific social group could be identified as a 'social fragment' but may not coincide in its residence or everyday fields of action with a particular 'physical fragment' (Vidal, 1999, pp. 159–169). On this basis, 'fragmentation' is the underlying condition of the built environment that opens further reflections around their relationality. If we take Vidal's process-oriented view of fragmentation as the multiscalar production of physical units that interact through a more or less integrated spatial lattice, then the interfragmentary space of interaction constitutes the realm of the interstitial geography: a 'field of relations where transitions between fragments are produced' (Vidal, 2002, p. 147). The production of 'fragments' supposes the production of an 'interfragmentary space' of relations, although we still do not know how these interfragmentary spaces are and what their functions are. According to Vidal, urban fragments are composed of a 'core' surrounded by a 'field', which is connected to the core. Yet, the 'interfragmentary space' is outside the fragment's field. Thereby, the interfragmentary space appears as a 'zone of reconciliation' which permits explicit recognition of an urban fragment's borderlands (Vidal, 2002).

On a wider scale, this idea of fragments, a surrounding field, and an inter-fragmentary space of interaction also connects to the notion 'Functional Urban Area' (FUA) found in the EU-OECD policy. The policy indicates that the FUA is composed of an urban centre, the city, and its commuting zone. The resulting FUA would be the combination of the city and its commuting zone (Dijkstra et al., 2019). This definition instrumentalises the surrounding area of cities by incorporating critical demographics in the work-residence relation. The FUA becomes a key for the emergence of the interstitial space when there is a regional interaction through the space comprehended between two cities or urbanised regions. This is described as the 'polyFUA' – the next level of functional consistency – that appears 'when contiguous FUAs are merged' (Meijers & Burger, 2017, p. 277). Finally, metropolitan regions that are less functionally integrated describe an in-between space of potential integration named 'Potential Integration Areas' (PIAs). The PIA emerges by merging FUAs and defines the potential relationality of regional interstitial spaces. A distinction is made between isolated cities and those positioned near other cities as cases of multicentricity. These cities are located in an FUA, polyFUA, or PIA that includes multiple cities (Figure 3.1).

3.4 Towards a definition of the interstitial space

Although the notion of 'interstitial space' is used in architecture and urban design (Steele & Keys, 2015; Sankalia, 2010; Verderber & Fine, 2000; Vondrak & Riley, 2005) – its translation into urban studies has been inciden-tal, random, somehow ambiguous or with a strong focus in only one aspect (economic, social, spatial, others) of its multifaceted nature. It seems that the meaning of interstitial space somehow 'goes without saying', a similar assumption that can be made for other critical terms such as 'place', 'space', 'city', 'infrastructure', or 'region' (to mention some). However, these terms have been more carefully used once further scrutinised or used to describe a certain type of elements with a very specific and clear empirical tractability; so, the same should apply for the term 'interstitial space'.

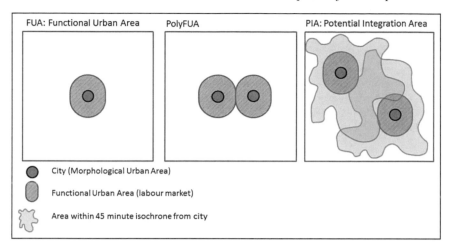

Figure 3.1 Meijers and Burger's FUA, polyFUA, and PIA.

Source: Reproduced with the permission of Meijers and Burger (2017).

In particular, it is argued that the term 'interstitial space' is more embracing as it encapsulates the varied and multifaceted elements of the interstitial geography and appears as suitable to operationalise the analysis of these varied elements, spaces, places, and functions within, between, and beyond sprawling cities and regions. In some way, the level of abstraction suggested by the notion itself apparently makes it less tractable in empirical terms when talking about cities. While in architecture, an 'interstitial space' can be a very concrete space between two or more buildings (considering that buildings have clear physical boundaries), an 'interstitial space' at an urban and regional scale can be more diffuse. What may be an interstitial space at the scale of sprawling cities and regions? What may it be if not something that is simply in-between something else? Could it be a 'space' or a 'place'?

The generic meaning of 'interstitial space', however, is its main strength. A closer inspection of the term reveals that its abstract character is not such when dissecting its ontological meanings and the interdisciplinary conceptual commonalities found in the vast fields of social sciences, arts, engineering, geography, urban studies, and spatial planning. The fact that a battery of terms is needed to differentiate the different expressions of the interstitial spaces as in-between non-spaces/places, undeveloped, under-developed, sub-developed, leftovers, etc. only unveils their varied origins and empirical expressions while demonstrating their multifaceted morphologies, functions, the scale at which they manifest, their different temporalities, relationalities, and potentials. In that sense, the 'interstitial space' better encapsulates this variety and embraces the nature of these different spaces as 'places' and 'non-places'.

The conceptualisation of the term 'interstitial space' is necessary to understand its substantial meanings but mainly its empirical tractability in the broad field of urban studies. This exercise is done by using an ontological and linguistic approach (Kim et al., 2019; Frisson et al., 2008) in which the conceptual meanings of 'interstitial' and 'space' are dissected separately to then be reformulated as a single conceptual construction that has both abstract and empirical significance. The term 'interstitial space' is made up of the words *interstitial* and *space*, and as such, will be treated as a linguistic assumption in which the two independent nouns (*interstitial*, and *space*) are re-signified to be then used together as an open compound that signifies only one final noun: *interstitial space*. It is proposed as an open compound as the two terms show a certain degree of opacity – ergo need to be identified – while together become sufficiently known to rely on its high level of semantic transparency.

3.4.1 Dissecting the notion of 'interstice'

The term 'interstitial' has a Latin root, which partly explains why it has been hardly used in the Anglo-Saxon literature while commonly used in both the specialised and non-specialised literature in Spanish and French inter alia. As such, it comes from the Latin *'interstitium',* in which *'inter'* means 'between', and 'stitium' means 'site' or 'place'. So, 'interstitial' would refer to 'a place (or site) in between' (Lewis & Short, 1879). Over the years, however, the term has become one in itself, mainly referring to an 'in-between' condition while losing its association to the idea of 'place' or 'site'. This is why several 'conceptual surnames' – in the form of attached words – accompany the word 'interstice' to better specify what this refers to. This 'conceptual surname' makes the 'interstice' empirically tractable while referring to something observable. Thereby, it is possible to find – in several disciplines – the terms 'interstitial alloy', 'interstitial compound', 'interstitial element', 'interstitial fluid' and 'interstitial space' to mention some. The latter makes sense to be used in urban studies considering that not all the places that lie between (and beyond) the built-up geography of sprawling cities and regions can be properly considered 'places' as such.

Etymologically speaking, there is a consensus that the term 'interstitial' does not automatically refer to a 'place' or 'space', but to any 'entity' – that can be a space, a physical element, or an interval of time – situated between two or more entities that can also be spatial, physical or temporary. Therefore, it is possible to identify an 'interstitial element' that is not necessarily a place or a space but a solid (and eventually inert) object between two or more solid objects. The association of 'interstice' to 'a space', however, is also gaining traction while talking about the small or narrow distance between elements. According to The American Heritage-Dictionary of the English Language (2000), an 'interstice' is 'a space, especially a small or narrow one, between things or parts'. Similarly, the Collins English Dictionary

(2014) an 'interstice' is 'a minute opening or crevice between things'. The American Heritage Science Dictionary (2005) indicates that an 'interstice' is 'an opening or space, especially a small or narrow one between mineral grains in a rock or within sediments or soil'.

In Physics and Chemistry, an 'interstice' describes special elements within the matter. This conception has been used since the 1950s in studies on atomic energy, minerals, and chemical reactions. In these fields, an 'interstitial compound' or 'interstitial alloy' is an element that appears when an atom – of sufficiently small radius – sits in an 'interstitial hole' of a metal lattice and thus, becomes an 'interstitial element' (Wells, 1962). This 'interstice' is a sort of 'mistake' or 'impurity' in the fabric of pure elements. According to Plumtree and Gullberg (1980), an 'interstitial element' is 'an impurity' found in pure metals, usually introduced during the manufacturing process (Plumtree & Gullberg, 1980). In biology and medical sciences, the term 'interstice' describes small spaces within organic structures. The literature identifies 'interstitial inflammations', 'interstitial fibrosis', 'interstitial lung diseases', 'interstitial pregnancy', the 'interstitial tissue', 'interstitial cells', the 'interstitial compartment', 'interstitial fluids' and others (Fleischhauer et al., 1995; Kaissling et al., 1996). In Arts, an 'interstice' is a 'conceptual gap' between well-known artistic categories or a superposition of several artistic styles. It could be a hybrid style resulting from the mix of other artistic genres such 'fiction' and 'horror' (horror-fiction); an artistic genre that falls 'in-between' rather than 'within' previously accepted categories (like 'interfiction', which is interstitial fiction) (Soyka, 2007). In Computing Sciences and studies about the Internet (or the 'inter'-'net'), the 'interstitials' are small websites displayed while the main website is being loaded (Jansen, 2002). These 'interstitials' display advertisements or requires confirmation of data before moving on to the next website ('hyperstitials'). In theological studies – and specifically in Roman Catholicism – an 'interstice' is a temporary stage of transition; 'a period of three months between the diaconate and the priesthood ordination' (Hardon, 2007).

As an intrinsically 'in-between' condition, an 'interstice' supposes the inevitable presence of surroundings or boundaries (in *between what?*). Thus, the 'interstice' supposes the existence of a context, a system of boundary objects – or surroundings – that configure the interstitial condition as such. Therefore, if the boundary objects do not exist, it follows that there is no interstitial (in-between) entity either. An 'interstice' is context-dependent – in the sense that it only emerges (or exists) insofar there are boundaries – and would not exist if the boundaries disappear. It is also possible to infer that the boundaries need to be identifiable or distinguishable from the interstice itself, contrasted by, or different from. A portion of 'space' – in the sense of a void, a vacuum, or an absolute empty space – cannot be recognised as 'between' other empty spaces if the boundaries cannot be identified. This would literally be 'a space between spaces' that would be more an abstract construction with no interstitiality at all. This is why it is not possible to recognise an apartment space in a housing block if there are no floors, doors, or walls that separate one apartment from another. Equally, it is not possible to recognise an (interstitial) 'valley' if there are not surrounding mountains.

3.4.2 Scrutinising the notion of 'space'

With no intention of reducing the number of reflections and profound insights built upon the notion of 'space', it is possible to assume that in the last decades, at least three critical thinkers have undoubtedly been influential in their treatment of the notion of space applied to the reality of cities and spatial thinking. The seminal work of Michael Foucault is foundational in the understanding of the urban spaces and their relations with human behaviour. In a second instance, the distinctions between 'space', 'place' and 'non-place' proposed by Marc Augé are fundamental to comprehend the varied (and empirically tractable) spaces of networks that talk about different mobilities and temporalities. Finally, the influential ideas of Henry Lefebvre are necessary to connect the political meanings associated with the production (and reproduction) of the space and explore how they connect to the production of 'interstitial spaces'.

Foucault's theoretical corpus has to a certain extent informed the study of space in the contemporary world. In particular, his concept of 'heterotopia' encapsulates some of the fundamental characteristics of the 'space' as a construction for the 'interstitial'. In his historical revision of 'the space', Foucault expands the (static and terrestrial) medieval notion of space as 'emplacement' (or location) of hierarchical groups that take position in a portion of land, to a conscious cognition of the space as a dynamic field of extension, where there are no fixed places; everything becomes a transitioning (and evolving) point (or instance) in movement. Ultimately, Foucault's view of 'space' in the present era is one in which space takes the form of relations among sites (...) defined by relations of proximity between points or elements (Foucault, 1984). This highlights the relationality of the space as an attribute of the 'interstice' in the sense of bridging other spaces (points or elements) while 'in-between' them. Another point relates to Foucault's view of contemporary society as a system of simultaneous events; processes of juxtaposition and dispersion occurring at the same time, and in which the space acts as an elastic platform for simultaneous social processes: 'a network that connects points and intersects with its own skein' (ibid). Under this view, the space is not an absolute entity but a dynamic one placed in a heterogeneous web of divergent spaces.

By using the metaphoric 'heterotopia' (ibid), Foucault highlights the un-ideal situation of an element in a new position, space, or place, in which its functions are modified or undermined. An element – that was designed, planned, or conceived for a certain function in a determined context – placed in a new space needs to be adapted, re-planned, re-designed, or re-programmed to become consistent with its new context. This is the opposite to a 'utopia' – also used as a contrasting tool by Foucault – as 'whereas utopias are unreal, fantastic, and perfected spaces, heterotopias in Foucault's conception are real places that exist like "counter-sites", simultaneously representing, contesting, and inverting all other conventional sites' (Sudradjat, 2012, p. 29). In that sense, if cities and the built-up space are somehow the outcome of plans and design – although not completely 'ideal' – and thus, more closely attached to the 'utopian' approach, the interstitial spaces are 'heterotopic' in nature while composed of elements not fully planned or organised to be part of. The 'heterotopic' character of the interstitial space presents juxtapositions of institutional arrangements and governance, differing functions, land-uses, and spatial relations, a context that represents incompatible spaces and reveals paradoxes; a context that receives rejected artefacts from the (idealised) city such as old industries and obsolete energy plants expelled to be relocated. The 'interstitial space' is an entropic context in which the accumulation of 'heterotopias of deviance' and 'crisis' intensify disorder, disorganisation, and dispersion. The interstitial space is heterotopic in nature because of its degree of economic and social alienation from urban plans, regional policies, and the design of cities; it has no inherent order and no rational connections between its parts. The interstitial space's heterotopia results from processes of change and hybridisation facilitated and built around a variety of patches and enclaves that are interconnected by an almost unlimited set of interfering networks (physical and virtual) (Shane, 2005). This heterotopic nature accounts for the incompatible and contrasting representations of the urban – meaning as spaces to pass through, to stay, and to live in – also lawlessness, chaotic, polluted, anti-social, without obvious mechanisms of overall coordination, and sources of radical environmental changes.

Augé's arguments about the proliferation of 'non-places' in contemporary society have had a significant impact across the humanities and social sciences, surfacing in such disciplines as geography, cultural studies, urban design, sociology, performance studies, architectural theory, contemporary philosophy, and literature inter alia. In his seminal piece 'Non-Places; Introduction to an anthropology of supermodernity' (1995), Marc Augé examines the changing characteristics of 'space', 'place' and individuality in a (super)modern world characterised by the acceleration of processes of production, the proliferation of simultaneous events, sped-up of information flows, the acceleration of history, movement, and the continuous abstraction of the space.

Augé argues that in the 'supermodernity' there is a dual production of spaces: 'places' and 'non-places'; entities that are rather like opposed polarities, although simultaneously coexist and mutually reconstitute themselves in a dialectical relation where 'the first is never completely erased, the second never totally completed' (Augé, 1995, pp. 78–79). For Augé, the 'space' becomes a (anthropological) 'place' when familiar, localised, historic, organic, meaningful social codes and identities flourish. In 'a place', the societal processes are strongly connected to the space where these processes and histories take place. This 'place' might be exemplified through a historic village or town which appears to have a strong sense of local identity, sense of community and attachment, and a geography and history that is (more or less) shared by the individuals that compose the community. This is the typical situation of cities in which – over the years – different layers of construction, governance, social participation, consolidation of collective memory, cultural production, and economic and historical processes have shaped the urban form and crystallised its subjacent social meanings over time.

By contrast, Augé examines the emergence of 'non-places' when the 'space' becomes an abstract ad standardised entity built upon mediatised images at the point in which individuals becomes the witness of contemporary life (consumers of the space) rather than active actors of its societal processes (producers of space). This spatialised experience is the 'non-places': spaces of circulation, communication, consumption, virtuality, endless transit, spaces 'where solitudes coexist without creating any social bond or even a social emotion' (Augé, 1996, p. 178). The 'non-place' is a space-time instance characterised by a partial or total detachment between the individual and the space; a reaction against place-attachment and place-identity in which both – the space and the individual – become anonymous. In the 'non-place', the loss of meaning and loss of proper connection between locations, the geographies of 'otherness' and 'nowhereness' become the norm (Arefi, 1999). As such, the 'non-place' is charged by information and signs that connect the individual with transnational codes of life and history – a sort of alienation from the proper place – by transporting its consciousness to other spaces, identities, and ideals. At some point, a 'non-place' becomes a neutral space for 'the average' person; ubiquitous spaces of temporary dwellings or communities – with no local cultural roots of any sort – that amalgamate secondary homes and neighbourhoods, temporary accommodations for workers, airports, motorways, hotels, shopping centres, theme parks, cyberspace, and tourist spaces. The 'non-places' have a material form and morphology that corresponds to the infrastructures of communication, transport networks, and enclaves of capitalist exploitation. These are the 'empirical non-places' that Augé suggests are being continually expanded as part of the decentring process of supermodernity; archetypical structures of what Edward Relph (1976) remarked upon the proliferation of motorways, railways, and airports, which constituted both forces and symptoms of 'placelessness'. If we take Augé's anthropological conception of the space as a 'non-place' – in the sense of being functional to transitional stages of movement, ephemeral temporalities, trans-spatiality, socio-cultural collective detachment – then the interstitial spaces between cities and regions need to be considered an integral example of in-between 'non-places'.

Arguably one of the most influential works in the field of urban studies, Lefebvre's seminal book *'The Production of the Space'* (1991) has situated cities, the urban, and 'the space' at the core of political theory. It is important to notice that Lefebvre's work makes an important contribution not only to urban theory but, if appropriately understood, to theory more generally. Architects, designers, planners, and geographers (among others) have extensively used Lefebvre's concept of space as a social product to frame urban and political critique. By conceptualising 'space' as 'social space', Lefebvre moves the term away from its Cartesian references as he articulates social space through a primarily ontological engagement: a complex and multifaceted object that exists in three simultaneous but distinct, co-producing levels: the mental, social, and physical. There is still an open debate regarding the latent epistemological challenges of operationalising Lefebvre's triad, although Lefebvre's attempt to unite these three realms provides a comprehensive framework of the underlying politics behind the production of the space.

Lefebvre's hypothesis is that the space is not an inert entity in the sense that 'space' is not simply inherited from nature or autonomously determined by laws, norms, or technicalities of spatial geometry as per conventional location theory. Instead, Lefebvre's postulate is that 'humans create the space in which they make their lives; it is a project shaped by interests of classes, experts, the grassroots, and other contending forces' (Molotch, 1993, p. 887). The 'space' is 'produced' and re-produced through social processes that have political roots and economic salience. Therefore, 'production' implies that 'space' can be considered analogous to other economic goods and so commodified. This also implies that the 'space' is not a neutral place but an entity that serves specific interests in a relational scope of interchange, densities, infrastructures, and norms. These infrastructures – such as roads, buildings, strategic and commercial facilities – inevitably privilege certain kinds of activities and inhibit others in the sense of supporting projects of one type while deterring the goals of others. This situation talks about the relative relationality of 'the space'; a motorway in a suburban neighbourhood, for instance, would reflect regional (or national) interests while it can also interrupt the erstwhile relational potential of adjacent built spaces and communities inhabiting them. Moreover, the relational potential of the space exceeds the territorial purview by going beyond material impediments in terms of symbolic meanings connected to mental and social dimensions.

While produced, the space is disputed, contested, and (re)produced for those hegemonic groups that somehow place the underlying guidelines of how the space might be. These disputes in a global context depict a confrontation of diverse values, ideas, and preferred arrangements as 'people fight not only over a piece of turf, but about the sort of reality that it constitutes' (ibid, p. 888). The space represents those who are able to build a dominium through its recognisable morphology, infrastructures, and socio-political meanings associated to. These 'spaces of domination' (hence, Lefebvre's notion of 'abstract space' in the sense of serving abstract purposes such as the 'national interest') aim to more pervasively serve the reproduction of capital. At some point, the whole societal mechanism of production of the space – including the mobilisation of expertise and knowledge – is subdued to the production of these 'abstract spaces' of exchange of capital value (Vergara & Boano, 2020). The interstitial spaces are to be conquered, in the sense of what Lefebvre introduces as space of 'domination', a way of flagging a spatial domain that can transcend the material boundaries of normative site. These are the cases of large architectonic artefacts or mining (exploitation) sites, which reflect contemporary forms of accumulation and trans-national corporativism of global influence. These spaces depict the impetus of capitalist mainstream politics that touches all corners of new and pre-existing space while consolidating ideological pathways to ensure its own reproduction and eventually threatens nature itself.

Lefebvre's conceptualisation of the space as 'abstract' – opposed to the 'absolute space' – in some ways connects to Augé's notion of 'non-place'. While Lefebvre's absolute space is defined by the modes of life satisfaction linked to a 'logic of senses' (156) – including the harmonisation of the human body with the space, its perception, its kinetic experience moved by organic human needs and meaningful feelings – the interstitial spaces are the abstract ones. As such, they describe the dismembered dispersion of infrastructures, enclaves, agglomerations of different sorts and reflect the outcomes of intensified capitalism driven by empowered hegemonic groups. The apparently 'absence of unitary plan' in the making of infra-structures, satellite towns, large architectonic artefacts, privatised (gated) neighbourhoods, private natural reserves, farming lands, military facilities, technoburbs, and sites for exploitation of natural resources – all framed by corporative values – resonates with the heterogeneous nature of the intersti-tial geographies between cities and regions, an heterogeneity that is far from being morphologically consistent when compared with its counterpart, the sprawling (absolute) *built-up* (space).

3.4.3 *Finding the common ground*

Having depicted the conceptual structure and ontological interpretations of the 'interstice' – and its applications in different disciplines – it is possi-ble to identify that first, it generically refers to an 'in-between' condition. Second, the 'interstice' can be a space, a place, an element, or an inter-val of time contained 'in-between'. Third, as an 'in-between' condition it requires the existence of boundaries, surroundings, or a context that sustains the in-between condition as such. Fourth, the 'interstice' can be a separate or distinguishable entity that can be more or less (in)dependent on its surroundings. Finally – and if invoked to specify special functions, characteristics, or in-between entities – the 'interstice' would require a 'conceptual surname' that specifies its meanings and makes it tractable in empirical terms. In that sense, it would not be enough to say 'this is an *interstice*' while referring to a concrete (and observable) place, space, a temporary situation, or a solid element situated between other identifiable entities; it would be necessary, thus, to indicate that this is an 'interstitial space', an 'interstitial building', an 'interstitial city', an 'interstitial land-scape', etc.

Having scrutinised the 'space' from Foucault, Augé, and Lefebvre's contributions, it is possible to find some corresponding links in their helpful dual-based ontological exercises. Foucault's heterotopic space resonates with the space of disruption and entropy introduced by the 'heterotopias of deviance'; something found in Auge' non-places in the sense of detachment, endless transition and sense and anonymity, and also Lefebvre's abstract space of global capitalism and homogenisation. The infrastructures of the space – and its role as relational of different realities and devices of the urban – talk about its condition as an 'in-between' entity (between order and disorder, local and global, identity and anonymity, individual and collective, specific and standard...) and thus, interstitial. The 'space' that lies between the built-up, cities, and regions is therefore *interstitial* in nature and is a *space* in the sense of being a *heterotopic abstract non-place*, far from being a 'place' in the sense of utopian, absolute, or what Leitner et al. (2008), in following Lefebvre's ideas, describes as 'sites where people live, work and move, and where they form attachments, practice their relations with each other, and relate to the rest of the world' (161). It is important to notice that these insights are still under debate. Massey's (1991a, 1991b) early articulation of the 'global sense of place' stretches the scalar perspective on geographical consciousness and belonging, arguing that place-experience also depends on engagement, participation, involvement, and inference in decision-making – or the construction of a collective memory while 'in transit' – and not only pure localism. Since then, a series of related (though uncoordinated) efforts to better theorise the space and place as relational entities have emerged.

3.4.4 Situating the interstitial geography

Based on the meanings associated with the term 'interstitial space' in urban studies and its previously developed conceptual construction, it is proposed using the notion of '*interstitial geography*' to refer to the wider and multi-faceted spatial scope in which the interstitial spaces are manifested. The 'interstitial geography' would be composed of different 'interstitial spaces', 'networks', and 'interstitial hubs' that can (or cannot) be interconnected through the network space. The latter – the 'interstitial hubs' – would empirically manifest in the form of dispersed buildings, isolated neighbourhoods, large retail facilities, and other architectonic artefacts, infrastructures, agglomerations, and enclaves of different sorts, villages, satellite towns, exploitation sites, gated landscapes, and others that can be considered as 'built-up' or as 'places', or well-established hubs of accumulation. As such, the notion of 'hub' would generically be used when its empirical tractability becomes ambiguous or uncertain. If not, it can be replaced by the well-known category to which it refers to; so, if an interstitial hub refers to a town located in an interstitial space between two cities, it would become an 'interstitial town'; if it refers to a mining site, it would become an 'interstitial enclave'; the same would apply to any landscape that can be identified as a clear accumulation hub that has taken a certain degree of Lefebvre's form of an 'absolute space'. In that case, it would be an 'interstitial landscape'. Although 'interstitial hubs' can be more or less interconnected with other neighbouring hubs, cities, or regions through the network space (virtual/physical), they can also transcend their locality to interact with the wider global space and other interstitial hubs located in other (far afield) interstitial spaces (Figure 3.2).

The interstitial hubs are key centres that generate both discrete and disproportionate amounts of economic growth and form the fabric of the interstitial spaces and the interstitial geography. The hubs talk about the space of networks that interconnect them and define their rationality, as well as the relational character of the interstitial spaces. They articulate the functional economic linkages with other hubs, industrial towns, and cities and highlight the importance of radical transitions between local and global economies; 'smaller functional economies that are otherwise marginalised and/or excluded by city-first agglomeration approaches' (Harrison & Heley, 2015, p. 1128).

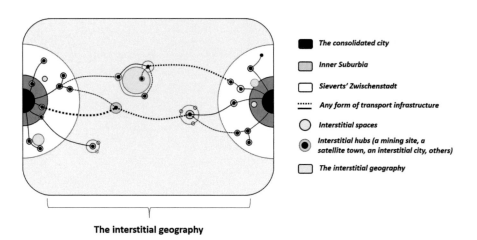

The interstitial geography

Figure 3.2 The interstitial geography and its components.

Source: Author.

It is important to emphasise that the sprawling space between (and beyond) cities and regions cannot be reduced to the rural scope. This amalgamation of interstitial spaces and hubs depicts a geography the suggests a new relationship between the city and the countryside (Massey, 2010). In many ways, capitalist accumulation is overflowing the different corners of urbanised regions and the interstitial spaces (Massey, 1991a). This process occurs in a relatively even manner in some regions and in a very disruptive one in others. In any case, capitalist accumulation penetrates over wider and wider swaths of countryside, becoming selective in terms of location and the scale in which it is manifested; from the city to the countryside, from the interstitial spaces through the interstitial hubs, different devices of commodification – including agriculture and nature itself – clear the way for the location of transitional functions of accumulation beyond the scope of cities. These functions range from cultural heritage, tourism and entertainment, agrotourism, and urban agriculture to other types of restructuring economic activities (Choy & Buxton, 2013). These penetrations have profound implications in the way of how we perceive the space and the conceptual apparatus that we use to analyse it. In the city, the notions of 'neighbourhood' and 'neighbours' take the form of a specific spatial proximity, while in the interstitial geography it can take the form of regional distances and spatial remoteness.

The 'interstitial geography' is the context in which these transformations occur; a context where emerging relations of 'city–interstitial geography' have made the rurality more visible; a space that is far from being homogeneous, monofunctional, and completely inert. The interstitial geography is spatially spotty, uneven, and is the space where emerging functionalities of environmental and economic transactions take place. This is the realm of the wicked problems; the context where some of the most radical economic changes and some of the most radical environmental alterations – inflicted upon the surface of the Earth – are reaching a point of no-return when linked to environmental justice and climate change (Goldman et al., 2018). The interstitial geography is a context where new approaches in innovation and creativity take place; approaches that are somehow unknown in the traditional space of cities; a context in which the disciplines of spatial planning, urban planning, urban design, and others – that have built their ethos around the built-up form – tend to collapse, or at least expose their precarious or limited hermeneutical tools to re-read the space. The interstitial geography suggests governance challenges around the elements of its fabric, while the institutions and socio-cultural processes that signify its politics are yet to be understood.

With a strong focus on contemporary transformations and semantics around the notion of 'countryside', Koolhaas' book '*Countryside, A Report (Countryside in your pocket!)*' (2021) also reclaims attention on the interstitial geography. While Koolhaas recognises that the term 'countryside' is a 'glaringly inadequate term for territory that is not urban' (3), he indicates that it includes rural areas but also oceans, beaches, and any non-strictly city-space. The book's central thesis is that current forms of urban life have necessitated the organisation, abstraction, and automation of the country-side at an unprecedented scale, although yet practically all attention goes on cities. Koolhaas illustrates that the countryside is subject to the same market forces encountered in cities, paying attention not only to the economics but also technological, environmental, and socio-cultural processes that are intensified between and beyond cities; processes in which leisure, work, large scale planning, political forces, climate change, migration, human and natural ecosystems, artificial intelligence and high-technology are altering the landscapes beyond cities across the world. The 'countryside' is a colossal interstitial geography organised with relentless Cartesian rigour, including advanced processes of data storage, fulfilment centres, robotics, business innovation, labour migration, private purchase of land for ecological preservation, and others. This heterotopic depiction of the countryside questions the widespread imaginary of a tranquil, safe, green, low-tech, and underused space. The (supposedly) pristine natural and farming environments across the Earth's vast non-urban areas are now a mix of genetic experiments, science, industrial nostalgia, seasonal immigration, large-scale land privatisation, massive subsidies, and tax incentives. The interstitial geography is also the context of forgotten (and under-siege) indigenous cultures, political turmoil, backyard of refugees ('altruistic segregation'), and pilot innovations on e-commerce. While in China entire rural villages are transformed into all-in-one factories or e-commerce stores, in the south of Chile a few foreign magnates buy entire natural ecosystems to be protected from the (exploiting) state (Hora, 2018); all elements of Foucault's 'heterotopias of deviance' and the sustain inattention resulting from premeditated politics of exploitation and spurious narratives of the interstitial as boring, empty and inert.

Theoretical entries to address the transformations of the interstitial geography also come from urban political ecology (UPE). Focusing on 'the urbanisation of nature', UPE would include the geographical imprints of expanding global metabolic flows of matter, energy, and capital developed by some interstitial enclaves (Kaika & Swyngedouw, 2000). It has been demonstrated that many interstitial towns and communities closer to mining sites have been overburdened with massive energy undertakings that power mining operations that supply raw materials to international markets (Arboleda, 2016). Somehow, these towns are supportive of the global urbanisation process through their tribute to globally up-scaled infrastructures and more powerful interstitial agglomerations. Mining sites, trans-national retail businesses, high-tech hubs, and extended agricultural plantations are colossal landscapes that, while connected globally, describe the most radical alterations to the environment. Their sheer magnitudes alert us that the metabolic exchanges of matter, energy, and capital that feed the urban world are distorted and up-scaled to the point that they have now reached a hypertrophic, global extent. In parallel, some smaller, liminal, and discrete interstitial spaces and hubs – in the form of traditional structures of the countryside such as satellite towns or cross-road villages – are being boosted or disappearing as part of processes of borrowing size and growth shadow effects (Meijers and Burger, 2017).

The dynamics and transformations of the interstitial geography through the network spaces have been captured by Massey and Featherstone (2009) while indicating that the urban is made up of bundles or constellations of the space–time trajectories of various kinds of elements: human capital and networks of social relations, physical objects including rocks or mountains, the built environment of cities, etc. These objects are clustered (in planned or unplanned ways) by interactions of institutions and individuals at multiple scales. The shape and contents of those clusters are constantly socially and politically negotiated. It means that the social experience of the interstitial hubs as places is concrete, while these are socio-politically produced (Pierce et al., 2011). This duality of the space (experience/production) is the one that captures the different temporalities of the interstitial spaces, their nature as planned and unplanned spaces, the scale at which they are manifested, and their relationality.

3.5 Conclusions

In this chapter, the emergence of interstitial spaces has been clarified. This demonstrates that interstices are not mere leftovers of the urbanisation process but outcomes of intermixed underlying politics of locality and capitalist globalisation. The different and interlinked determinants that explain how and why interstitial spaces are produced demonstrate their condition as elements subdued to normative restrictions similar to those that shape the built environment. The different terminologies around the interstitial spaces also illustrate their varied and multifaceted nature while referring to very specific aspects of the interstitial condition as such. This allowed finding the commonalities to ground the inherent condition of the interstices as 'in-between' entities.

The exercise of dissecting the terms 'interstitial' and 'space' allowed the inspection of their meanings in conceptual and empirical terms. It is proposed that the composed term 'interstitial space' better encapsulates the multifaceted character of the elements that compose the non-urban geographies between the built-up areas, cities, and regions. The interstitial spaces must be understood as an integral expression of Foucault's space; while cities become familial (and somehow predictable) environments, the interstitial spaces are deeply unknown and heterotopic. This heterotopia connects to Augé's transitional 'non-places' resulting from extended processes of globalised 'supermodernity', to define the intrinsic nature of interstices as 'spaces' in the sense of Lefebvre's abstract space. The interstitial spaces finally construct a wider scope of relationality – the interstitial geography – where different infrastructures, networks, and interstitial hubs shape the interstitial geography, the expanded fabric of Sievert's *Zwischenstadt*.

References

Alvey, A. (2006). Promoting and preserving biodiversity in the urban forest. *Urban Forestry & Urban Greening*, *5*(4), 195–201.

Arboleda, M. (2016). In the nature of the non-city: Expanded infrastructural networks and the political ecology of planetary urbanisation. *Antipode*, *48*(2), 233–251.

Arefi, M. (1999). Non-place and placelessness as narratives of loss: Rethinking the notion of place. *Journal of Urban Design*, *4*(2), 179–193.

Augé, M. (1995). *Non-places. Introduction to an anthropology of supermodernity*. Verso.

Augé, M. (1996). Paris and the ethnography of the contemporary world. In M. Sheringham (Ed.), Parisian Fields (pp. 175–181). Reaktion Books Ltd.

Aurora, A. L., Simpson, T. R., Small, M. F., & Bender, K. C. (2009). Toward increasing avian diversity: Urban wildscapes programs. *Urban Ecosystems*, *12*(3), 347–358.

Barkasi, A., Dadio, S., Losco, R., & Shuster, W. (2012). Urban soils and vacant land as stormwater resources. In Eric D. Loucks (Ed.), *World environmental and water resources congress 2012* (pp. 569–579). American Society of Civil Engineers.

Batty, M. (2016). Empty buildings, shrinking cities and ghost towns. *Environment and Planning B: Urban Analytics and City Science*, *43*, 3–6.

Batty, M. (2016). Empty buildings, shrinking cities and ghost towns. *Cities*, *43*, 3–6.

Berger, A. (2006). *Drosscape: Wasting land urban America*. Princeton Architectural Press.

Brighenti, A. M. (2013). *Urban interstices: The aesthetics and the politics of the in-between*. Ashgate.

Bruinsma, F., Pepping, G., & Rietveld, P. (1993). *Infrastructure and urban development; the case of the Amsterdam orbital motorway*. Serie research memoranda. Faculteit der Economische Wetenschappen en Econometrie. Vrije Universiteit Amsterdam.

Casey, E. (1993). *Getting back into place. Toward as renewed understanding of the place-world*. Indiana University Press.

Chiesura, A. (2004). The role of urban parks for the sustainable city. *Landscape in Urban Planning*, *68*(1), 129–138.

Choy, D. L., & Buxton, M. (2013). Farming the City Fringe: Dilemmas for peri-urban planning. In Q. Farmar-Bowers, V. Higgins, & J. Millar (Eds.), *Food security in Australia*. Springer.

Clawson, M. (1962). Urban sprawl and speculation in suburban land. *Land Economics*, *38*(2), 99–111.

Collins English Dictionary – Complete and Unabridged Edition (2014). 12th edition. HarperCollins Publishers.

de Solá-Morales, I. (2002). Territorios. Editorial Gustavo Gili.

Dijkstra, L., Poelman, H., & Veneri, P. (2019). The EU-OECD definition of a functional urban area. (OECD Regional Development Working Papers, No. 2019/11). OECD Publishing, 1–18. https://doi.org/10.1787/d58cb34d-en (accessed May 2021).

Dovey, K. (2012). Informal urbanism and complex adaptive assemblage. *International Development Planning Review*, *34*(4), 349–367.

Fleischhauer, J., Lehmann, L., & Kléber, A. (1995). Electrical resistances of inter-stitial and microvascular space as determinants of the extracellular electrical field and velocity of propagation in ventricular myocardium. *Circulation*, *92*(3), 587–594.

Foo, K., Martin, D., Wool, C., & Polsky, C. (2014). The production of urban vacant land: Relational placemaking in Boston, MA neighborhoods. *Cities*, *35*, 156–163.

Foucault, M. (1984). Space, knowledge, and power. In P. Rabinow (Ed.), *The Foucault reader. An introduction to Foucault's thought* (pp. 239–256). Pantheon Books.

Frank, A., & Stevens, Q. (2006). *Loose space. Possibility and diversity in urban life*. Routledge.

Frisson, S., Niswander-Klement, E., & Pollatsek, A. (2008). The role of semantic transparency in the processing of English compound words. *British Journal of Psychology*, *99*(1), 87–107.

Gandy, M. (2011). Interstitial landscapes: Reflections on a Berlin corner. In M. Gandy (Ed.), *Urban constellations* (pp. 149–152). Jovis.

Gandy, M. (2013). Marginalia: Aesthetics, ecology, and urban wastelands. *Annals of the Association of American Geographers*, *103*(6), 1301–1316.

Goldman, M. J., Turner, M. D., & Daly, M. (2018). A critical political ecology of human dimensions of climate change: Epistemology, ontology, and ethics. *WIREs Climate Change*, *9*(4), 1–15.

Graham, S. (2000). Constructing premium network spaces: Reflections on infrastruc-ture, networks and contemporary urban development. *International Journal of Urban and Regional Research*, *24*(1), 183–200.

Hardon, J. (2007). *Modern Catholic dictionary*. Eternal Life Publications.

Harrison, J., & Heley, J. (2015). Governing beyond the metropolis: Placing the rural in city-region development. *Urban Studies*, *52*(6), 1113–1133.

Hora, B. (2018). Private protection initiatives in mountain areas of Southern Chile and their perceived impact on local development – The case of Pumalín Park. *Sustainability*, *10*(5), 1584.

Hugo, J., & du Plessis, C. (2020). A quantitative analysis of interstitial spaces to improve climate change resilience in Southern African cities. *Climate and Development*, *12*(7), 591–599.

Ige, J. O., & Atanda, T. A. (2013). Urban vacant land and spatial chaos in Ogbomoso North local government, Oyo State, Nigeria. *Global Journal of Human Social Science & Environmental Science & Disaster Management*, *13*(2), 28–36.

Jansen, E. (2002). *NetLingo. The internet dictionary* (p. 206). NetLingo Inc.

Jorgensen, A., & Keenan, R. (2012). *Urban wildscapes*. Routledge.

Kaika, M., & Swyngedouw, E. (2000). Fetishizing the modern city: The phantasma-goria of urban technological networks. *International Journal of Urban and Regional Research*, *24*(1), 120–138.

Kaissling, B., Hegyi, I., Loffing, J., & Le Hir, M. (1996). Morphology of interstitial cells in the healthy kidney. *Anatomy and Embryology*, *193*(4), 303–318.

Kim, S. Y., Yap, M. J., & Goh, W. D. (2019). The role of semantic transparency in visual word recognition of compound words: A megastudy approach. *Behavior Research Methods*, *51*(6), 2722–2732.

Kitha, J., & Lyth, A. (2011). Urban wildscapes and green spaces in Mombasa and their potential contribution to climate change adaptation and mitigation. *Environment and Urbanization*, *23*(1), 251–265.

Koohsari, M. J., Mavoa, S., Villanueva, K., Sugiyama, T., Badland, H., Kaczynski, A. T., & Giles-Corti, B. (2015). Public open space, physical activity, urban design and public health: Concepts, methods and research agenda. *Health & Place, 33*, 75–82.

Koolhaas, R. (2021). *Countryside. A report. The countryside in your pocket!*. Taschen.

Koolhaas, R. (1995). *Imagining the Nothingness*. In R. Koolhaas & B. Mau (Eds.), *S, X, L, XL*. The Monacelli Press.

La Greca, P. L., Rosa, D., Martinico, F., & Privitera, R. (2011). Agricultural and green infrastructures: The role of non-urbanized areas for eco-sustainable planning in metropolitan region. *Environmental Pollution, 159*(01), 2193–2202.

Lefebvre, H. (1991). *The production of the space*. Basil Blackwell Ltd.

Leitner, H., Sheppard, E., & Sziarto, K. M. (2008). The spatialities of contentious politics. *Transactions of the Institute of British Geographers, 33*(2), 157–172.

Lewis, C., & Short, C. (1879). *A Latin dictionary*. Clarendon Press.

Loukaitou-Sideris, A. (1996). Cracks in the city: Addressing the constraints and potentials of urban design. *Journal of Urban Design, 1*(1), 91–103.

Manzo, L. (2003). Beyond house and haven: Toward a revisioning of emotional relationships with places. *Journal of Environmental Psychology, 23*(1), 47–61.

Massey, D. (June, 1991a). A global sense of place. *Marxism Today*. 24–29.

Massey, D. (1991b). The political place of locality studies. *Environment and Planning A, 23*, 267–281.

Massey, D. (2010). A global sense of place (pp. 9–26). Aughty.org. http://banmarchive. org.uk/collections/mt/pdf/91_06_24.pdf (accessed in April 2021).

Massey, D., & Featherstone, S. (2009). The possibilities of a politics of place beyond place? A conversation with Doreen Massey. *Scottish Geographical Journal, 125*(3–4), 401–420.

Meijers, E. J., & Burger, M. J. (2017). Stretching the concept of 'borrowed size'. *Urban Studies, 54*(1), 269–291.

Molotch, H. (1993). The space of Lefebvre. *Theory and Society, 22*(6), 887–895.

Monkkonen, P., Comandon, A., Escamilla, J. A. M., & Guerra, E. (2018). Urban sprawl and the growing geographic scale of segregation in Mexico, 1990–2010. *Habitat International, 73*, 89–95.

Northam, R. (1971). Vacant land in the American City. *Lands Economics, 47*(4), 345–355.

Pagano, M., & Bowman, A. (2000). Vacant land in cities: An urban Resource. The Brookings Institution – Survey series. Center on Urban & Metropolitan Policy. 1–9.

Pemberton, S., & Shaw, D. (2012). New forms of sub-regional governance and implications for rural areas: Evidence from England. *Planning Practice and Research, 27*(4), 441–458.

Pflieger, G., & Rozenblat, C. (2010). Introduction. Urban networks and Network Theory: The City as the connector of multiple networks. *Urban Studies, 47*(13), 2723–2735.

Phelps, N. A. (2017). *Interplaces: An economic geography of the inter-urban and international economies*. Oxford University Press.

Pierce, J., Martin, D. G., & Murphy, J. T. (2011). Relational place-making: The networked politics of place. *Transactions of the Institute of British Geographers, 36*(1), 54–70.

Plumtree, A., & Gullberg, R. (1980). Influence of interstitial and some substitutional alloying elements. In *Toughness of ferritic stainless steels* (pp. 34–55). American Society for Testing and Materials.

Prévôt-Schapira, M.-F., & Cattaneo, R. (2008). Buenos aires: La fragmentación en los intersticios de una sociedad polarizada. *EURE*, *34*(103), 73–92.

Relph, E. (1976). *Place and placelessness*. Pion.

Rickards, L., Gleeson, B., & Boyle, M., et al. (2016). Urban studies after the age of the city. *Urban Studies*, 53(8): 1523–1541.

Sankalia, T. (2010). Kevin Lynch, Walter Benjamin and interstitial space in San Francisco. Department of Art and Architecture, University of San Francisco, 1–17.

Saunders, J. (2011). Recreational and ecological politics in the in-between City: The ongoing development of Downsview Parks. In D. Young, P. Wood, & R. Keil (Eds.), *In-between infrastructure: Urban connectivity in an age of vulnerability* (pp. 237–249). Praxis (e)Press.

Savarda, J., Clergeaub, P., & Mennechezb, G. (2000). Biodiversity concepts and urban ecosystems. *Landscape and Urban Planning*, *48*(3–4), 131–142.

Sayın, Ö, Hoyler, M., & Harrison, J. (2020). Doing comparative urbanism differently: Conjunctural cities and the stress-testing of urban theory. *Urban Studies*, 1–18.

Schnell, I., & Benjamini, Y. (2005). Globalisation and the structure of urban social space: The lesson from Tel Aviv. *Urban Studies*, *42*(13), 2489–2510.

Shane, D. (2005). *Recombinant urbanism: Conceptual modeling in architecture, urban design, and city theory*. John Wiley & Sons Ltd.

Shaw, P., & Hudson, J. (2009) The qualities of informal space: (Re) appropriation within the informal, interstitial spaces of the city. *Proceedings of the Conference Occupation: Negotiations with Constructed Space* (pp. 1–13). University of Brighton.

Sieverts, T. (2003). *cities without cities. An interpretation of the Zwischenstadt*. Spon Press.

Sousa-Matos, R. (2009). Urban landscape: Interstitial spaces. *Landscape Review*, *13*(1), 61–71.

Soyka, D. (2007). Interfictions, edited by Theodora Goss and Delia Sherman. *Strange Horizons*. At: http://strangehorizons.com/non-fiction/reviews/interfictions-edited-by-theodora-goss-and-delia-sherman/ (accessed in May 2021).

Stamps, A., & Smith, S. (2002). Environmental enclosure in urban settings. *Environment and Behavior*, *34*(1), 781–794.

Stavrides, S. (2007). Heterotopias and the experience of porous urban space. In K. A. Frank, & Q. Stevens (Eds.), *Loose space: Possibility and diversity in urban life* (pp. 174–192). Routledge.

Steele, W., & Keys, C. (2015). Interstitial space and everyday housing practices. *Housing, Theory and Society*, *32*(1), 112–125.

Sudradjat, I. (2012). Foucault, the other spaces, and human behaviour. *Procedia-Social and Behavioral Sciences*, *36*, 28–34.

The American Heritage science dictionary (1st ed.). (2005). Houghton Miffin Company.

The American Heritage® dictionary of the English language (4th ed.). ([2009] 2000). Houghton Mifflin Company.

Theobald, D. M. (2001). Land-use dynamics beyond the American urban fringe. *Geographical Review*, *91*(3), 544–564.

Trancik, R. (1986). *Finding lost space; theories of urban design*. Van Nostrand Reinhold Company.

Twerd, L., & Banaszak-Cibicka, W. (2019). Wastelands: Their attractiveness and importance for preserving the diversity of wild bees in urban areas. *Journal of Insect Conservation*, *23*(3), 573–588.

Verderber, S., & Fine, D. J. (2000). *Healthcare architecture in an era of radical transformation*. Yale University Press.

Vergara, F., & Boano, C. (2020). Exploring the contradiction in the ethos of urban practitioners under neoliberalism: A case study of housing production in Chile. *Journal of Planning Education and Research*, 1–5.

Vidal, R. (2002) Fragmentation de la Ville et nouveaux modes de composition urbaine. L'Harmattan.

Vidal, R. (1999). Fragmentos en tensión: Elementos para una teoría de la fragmentación urbana (Fragments in tensión: Elements for a theory of urban fragmentation). *Revista Geográfica de Valparaiso, 29/30*, 149–180.

Vondrak, S., & Riley, D. (2005). Interstitial space design in modern laboratories. *Journal of Architectural Engineering, 11*(2), 60–70.

Wells, A. F. (1962). *Structural inorganic chemistry* (3rd ed.). Oxford University Press.

Wolman, H., Galster, G., Hanson, R., Ratcliffe, M., Furdell, K., & Sarzynski, A. (2005). The fundamental challenge in measuring sprawl: Which land should be considered? *The Professional Geographer, 57*(1), 94–105.

Yuan, Z., Jian, K., & Hong, J. (2015). Study on restorative benefits of public open space in high-density city: Take Shenyang as an example. *Architectural Journal, 13*(5), 152–157.

Zhang, H., & Grydehøj, A. (2020). Locating the interstitial island: Integration of Zhoushan archipelago into the Yangtze River delta urban agglomeration. *Urban Studies*, 58(10) 2157–2173.

4 The nature of interstitial spaces

Today, even a 'new' city is familiar: a predictable accumulation of roads, towers, icons ... but as soon as we leave the urban condition behind us we confront newness and the profoundly unfamiliar.

(Koolhaas, 2021: 2)

4.1 Introduction

The interstitial spaces and hubs are perceived as unplanned entities resulting from the absence of normative restrictions beyond the realm of cities. The image of the informally occupied spaces between buildings and a countryside composed of disperse structures, lack of density, emptiness, and the coexistence of differing functions suggest a totally flexible space in terms of locational decisions, architectonic typologies, size of infrastructures, and controls on functionalities. This is a reasonable assumption considering that planning is associated with cities and serving the public interest – often translated into morphological, social, and infrastructural unity – in which any evidence of functional inconsistency and lack of sense of 'place' redound in an 'absence of planning' with ecological and social consequences (Rutherford, 2008).

Although from a planning perspective this would be a natural conclusion, it would not be precise to assume that mining sites, small towns and villages, regional parks, military and high-tech facilities, or the space comprehended between and outer airport and the city have not been subject to some sort of planning. Transport infrastructures are subject to several restrictions linked to speed, safety, and construction norms. A suburban power plant or a brownfield site cannot be assumed as originally 'unplanned' either. In fact, the supposedly unplanned landscape resulting from the arrival of the sprawling city that makes these facilities 'interstitials' turn the planned-unplanned reflection back to the city itself (*what reached what?*). Both – the city and the interstitial spaces – are equally subdued to the temporalities of transformation while 'pending' for more prodigious periods of economic growth. Their transformation can also be triggered by political and

DOI: 10.4324/9780429320019-4

environmental campaigns that make them undesirable, polluting, inhuman, or dangerous while included into narratives of smart-growth or urban sustainability derived into their demolition, retrofitting, or total abandonment.

In this chapter, the planned, unplanned, and temporary significance of the interstitial spaces is discussed. It touches the literature around the empirical grounds of the emergence of interstitial spaces and the role of planning (or 'what type of planning') in articulating the determinants of the interstitiality. It correlates to the interpretation of interstitial spaces as unplanned leftovers of less controlled processes in planning; as such, interstitial spaces are more the outcomes of the interaction of local and global forces of capitalist accumulation that places 'planning' in a stage of governance that overpasses the realm of cities and regions. The interstitial spaces articulate individual, social, and corporative motives that find in the interstitiality a blank canvas for experimentation and innovation beyond bi-dimensional conceptions of land-use. It proves that – although apparently unplanned and static – interstitial spaces are dynamic and 'pending spaces' in their transitions to eventually become something else. As such, the interstices result from different regimes of efficiency, integration, and appreciation of nature that run under similar analytical perspectives and normative agendas on urban dynamics found in the built environment. These aspects serve as the basis to understand the scale at which the interstitial spaces are manifested and their relational properties for continuous development, aspects that will be discussed in detail in the next chapters.

4.2 The empirical grounds of interstitial spaces

The policies around the production of interstitial spaces are varied and can be classified into those directly related to the production of built-up lands, the production of open and green space, and political rationales behind the production of the space. In all these approaches interstitial spaces are not random outcomes of less controlled processes in planning but produced alongside the urbanisation process; interstitial spaces evince that the politics around the built-up space equally influence their emergence (Phelps & Silva, 2017).

The spatial and institutional fragmentation that characterise urban sprawl have echoed with narratives of environmental sustainability and impacts on ecosystem services. There is an abundance of research from the gateways of political ecology (Heynen et al., 2006) and ecosystem services (Green et al., 2016) that shape the grounds against discontinuous environments. It is argued that land fragmentation is unsustainable for flora and fauna – and by extension – for humans. These critics are mainly based on the role of green spaces for wildlife – while brown, blue, and grey are often disregarded – in which physical barriers and the causes of the loss of biodiversity from an ecological perspective contribute to 'habitat fragmentation' (Said et al., 2016).

In a different vein, the ecological content of suburban land fragmentation also relates to the distribution of green space among segregated communities. The role of undeveloped areas, vacant lands, and abandoned spaces is valuable for politics of urban ecology and the study of socioecological systems formed from the combination of urban habitats for human beings and other organisms (Douglas, 2016). The implementation of different experiences of urban agriculture in interstitial spaces, for example, talks about the urban ecology of interstices while supporting the emergence of natural capital (Gandy, 2015) and interdependency of humans and natural systems (Wang et al., 2012). This interdependency is defined by the level of mutual accessibility to the different components of the environment (Lafortezza et al., 2013).

As part of the urbanisation process, it has clarified that although urban sprawl gives an 'unplanned impression', it is the sum of a series of individual (rational) decisions that have been throughout planned by different institutional levels and sectors, and in which interstices appear as part of (Sieverts, 2003). In this context, the environmental and social effects of sprawl 'have been as much a product of the state as of the private sector' (Phelps, 2015, p. 4), considering that public and private sectors have become barely distinguishable in modern (post)democratic societies (MacLeod, 2011). The question emerges, then, whether planning has produced the sprawling patterns of development – and if so, its interstices – or whether this is mostly a market-driven phenomenon. Whatever the case, what 'we can identify is the prevalence of "muddling through" as a mode of decision-making in relation to land-use and spatial planning' (Phelps, 2012, p. 13); an incremental process of urbanisation assessed case by case and in which interstices are also be embedded in.

The pace of this 'muddling through' has been historically different and so, the production of interstitial spaces. After WWII the building industry was fast enough to respond to demands around the 'suburban dream'. However, with the emergence of the anti-sprawl campaigns (Bruegmann, 2005), a growing 'pro-city' middle class – and the tactics raised by planning, architectural and design circles around suburban retrofitting – urban sprawl was slightly restrained. This simply gave space to more silence, piecemeal and scattered pattern of urban sprawl (Phelps, 2012) – consolidated by top-down approaches in planning – in which central authorities were reluctant to recognise their role as part of. As a scapegoat, they invoke the national interests while expanding the boundaries of local districts here and there: 'the norm – and in real life it is hard to conceive of an alternative – is incrementalism or "muddling through" in decision-making, including all forms of public policy' (Phelps, 2012, p. 5). The further promotion of sprawl usually takes place at weak locations found around motorway intersections, municipal boundaries, or where the repercussions on local politicians and authorities are minimised (Tewdwr-Jones, 1997). All of these historical processes of sprawling growth were built through the city-oriented focus, an approach through which urban sprawl has carried its negative connotation and where interstitial spaces became invisible.

As a way of contrast, the built-up space of the sprawling geography of many cities tends to be very homogeneous in comparison to the varied spatiality of the interstitial geography. While sprawling growth is often driven by a housing debate that derives into the implementation of homogeneous residential typologies, the interstitial geography is diverse and multifaceted. In some cases, the functions of interstitial spaces can be as significant as those found in iconic architectures. According to Home et al. (2010), from an anthropocentric perspective, the open landscape can provide similar services to those compared with a cathedral, as 'wilderness areas provide a vital opportunity for spiritual renewal, moral regeneration, and aesthetic delight' (Godfrey-Smith, 1979, p. 310 cited in Home et al., 2010, p. 496). In the Randstad area in the Netherlands, for instance, open spaces are considered significant 'icons' of the history of the country – recognised in the report *Icons of Dutch Spatial Planning* (Ministry of Infrastructure and the Environment, 2012) – in which at least 17 of these icons are natural open spaces and pieces of suburban countryside; valuable spaces for the region and whole society (Faludi, 2005). The morphological contrast between the built-up and the interstitial space is also partly explained by Lefebvre's hypothesis that capitalism directly influences practical conceptions of the 'space' by reproducing existing and socially accepted architectonic typologies. Innovation and difference for the built-up are restricted in 'the iterative role of technological innovation in the project of building, bounding, and subsequently managing the city' (Pierce & Martin, 2015, p. 1290), while interstitial spaces become non-standard. Any re-conception of urban forms linked to the transformation of the hegemonic socio-cultural realm – or the acceptance of emerging societal meanings – is rejected. So, the space is reproduced (again and again) by fulfilling market niches and by adapting designs to the rule of profit (Vergara & Boano, 2020).

The understanding of the specific determinants and the forces that operate upon the production of the 'interstitial space' can provide insights into the interdisciplinary nature of cities and the spaces that lie between developments. Critical voices from urban political ecology have indeed argued that these interstitial spaces need to be codified as par to the urbanisation process and thus, the disciplines of the built environment should expand their focus beyond urban agglomerations (Angelo & Wachsmuth, 2015). Such critique is particularly relevant for the understanding of Lefebvre's 'urban revolution' in the sense that the urban has produced a highly uneven spatial fabric composed of both urban and non-urban geographies (Wachsmuth, 2012). In this policy context, interstitial spaces are intrinsic components of sprawling development resulting from projects of modernisation placed at the core of neoliberal agendas that has reached global levels (Boland et al., 2017). Despite their apparently 'natural' (or spontaneous) emergence, they are also by-products of institutional interventions

(Phelps, 2015) and the traction of context-dependent determinants, far from being mere leftover of less controlled processes in planning.

4.3 The emergence of interstitial spaces

What explains the emergence of interstitial spaces is a combination of factors that involve processes of sprawling (sub)urbanisation with global salience, characterised by land fragmentation while producing built-up areas. This also includes the production of outer functional agglomerations – interstitial hubs – that lie within the regional interstitial spaces outside (and between) cities. Although these processes appear as randomly arranged – and as outcomes of development purposes – the production of the sprawling space and the interstices result from fragmentary actions over specific interests and planning restrictions (Phelps, 2012). As Sieverts indicates, the *Zwischenstadt* emerges from countless institutional and non-institutional (individual) decisions – each one with its own rationale (2003). In that sense, the production of interstitial spaces lies at the core of the same factors of the 'muddling through'; factors are varied and inter-linked. It would not be enough to indicate that interstitial spaces emerge as simple leftovers of fragmented extended (sub)urbanisation. Instead – and considering that the interstices are diverse in functions, morphologies, and environmental characteristics – their determinants are varied, interlinked, and manifest through different societal processes and politics.

In this book, the notion of 'determinant' is used to analyse the general causes and the specific factors that explain the emergence of interstitial spaces – including quantitative, qualitative, and context-dependent factors – as well as socio-political causes. Scrutinising the extant literature on interstitial spaces, the determinants can be classified into five categories: (a) *geographical and functional constraints*, (b) *socio-cultural*, (c) *economic*, (d) *policy-based*, and (e) *infrastructural*.

4.3.1 Geographical and functional constraints

In the early 1960s, studies on suburbanisation conducted in American cities identified areas in the countryside – including fertile lands, hills, and other geographical restrictions – as not suitable for new developments. Their physical constraints affected buildings costs, and thus, they remained undeveloped. Other situations included floodplains, landslides, and other physical handicaps that were instead considered to be integrated into suburban districts as open tracts; spaces that would socially benefit by providing leisure, recreation, parks, or natural reserves to be in contact with nature. Lands without accessibility and feasibility of water and energy supply were also on the list of remaining undeveloped (Clawson, 1962). A similar study published by Northam (1971) on urban lands in American cities identified that small parcels were usually not attractive due to their lack of land capacity for large infrastructure or real estate projects. Lands with irregular shapes also remained as leftovers as 'parcels that are extremely elongate, many-sided, constricted at one end, and the like do not lend themselves to usual forms of development, especially involving construction of a building' (p. 351). Similarly, Pagano and Bowman (2000) found that land size, plot shape, and location were the main factors of the interstitial condition of some lands. Using geographical analysis, Burchfield et al. (2006) found that physical (natural and infrastructural) barriers – and the environmental value of open space – are the main determinants of undeveloped lands. In several cases, the low cost of private water-wells facilitated the scattered concentration of houses in the rural space (Burchfield et al., 2006). Empirical studies conducted in Bandung – Indonesia – indicate that very thin soil layers or no/little water sources that make lands not productive for agriculture, forest, or even settlement define the emergence of undeveloped lands. Flood-prone areas that directly increase the risk of failure for real estate developments also determine the emergence of interstitial spaces. Vejre et al. (2007) – referring to the Copenhagen's Finger Plan – indicates that the protection of open space in the form of 'wedges' in many cases responded to a coincidence of preservation policies with the presence of natural restrictions such as lakes, undevelopable land, and old forests.

As part of the functional determinants, emerging agricultural industries in the peri-urban space also mean a restriction in transitioning farming land into built-up space. Although various policies and pressures have affected the continuous operation of farms in the urban fringe, productivist rural activities have transitioned to multi-functional (post-productivist) land-uses. It includes the preservation of lands with natural aesthetics and attracting new populations, protecting indigenous and cultural reserves, preventing forest exploitation, mining (including coal and gold), and attracting tourism (Choy & Buxton, 2013). Also, to minimise the impacts of urban sprawl farming lands tend to be preserved and included in plans as part of the environmental assets of the peri-urban space of city regions (Hong et al., 2017; Roche & Argent, 2015).

Interstitial spaces present important environmental properties within the whole process of urban expansion, but there are substantial contrasts in this role across a range that encompasses city centres, inner suburbs, outer suburbs, and peri-urban areas (Douglas, 2008). Most green spaces in inner city and inner suburban locations are protected and seldom changed (private gardens, private community squares, or small parks nearby). Interstices are planned as 'open' to sustain their environmental properties, create security zones, or create public spaces at different scales. These are the cases of green corridors, security buffers, parks, and squares. These 'vacant' lands are expected to provide positive externalities and often are secured under planning regulations. Some green spaces are also protected by plans aimed to diminish air pollution and noise and increase biological patrimony.

These lands are ecological entities that support ecosystem services at different levels (Kim et al., 2015) that 'may also be described as conservation or protection from extreme intervention such as imposed by development' (Maruani & Amit-Cohen, 2007). Larger ecologically active lands also tend to be protected. In the mid-Atlantic region, for instance, concerns about habitat fragmentation caused by the sprawling expansion derived into the preservation of forested habitats, largely recognised as beneficial for the regional vegetative community (Hess et al., 2001). Harvey and Works (2002) describe regulations for protecting these open (green) areas because of their importance in supporting micro-climate and environmental benefits also linked to the efficiency of local farmlands. Some of these regulations relate to the implementation of restrictions zones for urbanisation (such as the 'Urban Growth Boundary', UGB) to protect farms and forests from urban encroachments. The UGB of Portland was a state-mandated package of land-use laws explicitly aimed to protect non-urbanised areas (Harvey & Works, 2002).

4.3.2 The socio-cultural determinants

Permanent migrations to urban fringes also determine the presence of interstitial spaces. Since the creation of suburbs, the persistent mobility to suburban areas has been inspired by the proximity to nature, nature that needs to be preserved as part of the attractions of suburbia. New suburban developments consider open tracts as planned interstices that ensure contact with (controlled) nature. Families, children, elders, and visitors share these common open spaces in the reinforcement of their sense of community (Holcombe, 1999). It has been proved that in many European cities, planned and designed open space play an important role in local planning, which engage municipalities and multiple actors in partnership projects to attract new populations. Named 'discrete landscapes' (Santos, 2017, p. 7), local planning interventions consider interstitial spaces as part of a planned fragmentation of suburbia to promote a sense of community and ensure immaterial qualities of the space (ibid).

Interstitial spaces also emerge as a result of their historical significance. There are several spaces in the urban peripheries where archaeological ruins are found – many of them by accident while the sprawling urbanisation takes place – and are then preserved. These spaces are rich socio-cultural environments where ruins of the past coexist with the present city. In the case of Peru, for instance, Lima's peri-urban interstices define landscapes where desertic lands, millennial archaeological heritage (of the Inca empire), activism of local communities, informal occupations, nature and formal suburban structures are overlapped, questioning the character of interstices as environmentally inert spaces (Del Castillo & Sopla, 2018). It is argued that these spaces support the sustainable transformation of Lima's peripheries when they are integrated and codified in planning (Del

Castillo & Castro, 2015). In the case of Athens, several interstitial spaces were originally 'ancient places' (not interstitials) until the modern city arrived and surrounded them as part of its ruins. These interstitial spaces are preserved and transcend the local interest to become part of the global repository of historical heritage (Hill, 2013). In this category are also cemeteries often produced as semi-open public spaces in the form of parks and large squares. All of these interstices are symbolic spaces equipped with basic amenities to facilitate temporary visits or work (Li et al., 2014; Saxena & Sharma, 2013).

4.3.3 The economic determinants

In contexts of financial speculation, interstices appear as commodities or as public reservoirs for the expansion of services. Some interstitial spaces are acquired in advance – intended to remain undeveloped – to assume incremental demands of housing, services, and infrastructure. Northam (1971) found that some lands become interstitial through their role as 'corporate reserves': 'parcels of land owned usually by locally represented business corporations such as utility companies. The objective in owning such vacant parcels is to provide space for expansion' (Northam, 1971, p. 345). Not to be ignored is the vacant parcel held for speculation: '... parcels of land held in corporate owner-ship, in estates, or in single party ownership with the expectation that they will eventually be sold in the market place at which time a profit will be derived' (ibid, p. 346). In the case of African cities, Ige and Atanda (2013) demonstrated the correlation between the location, size (land-capacity), and value of land to determine whether it will become urbanised or remain interstitial (Ige & Atanda, 2013).

The emergence of interstitial spaces may be on the increase in contexts of urban shrinkage (Dubeaux & Sabot, 2018). Here, interstices appear as unused urban sites resulting from economic decay of previously active built-up areas. This is the creation of interstitiality 'from inside', in which previously active places start a 'hollowing' process resulting from the financial instability. It has been demonstrated that 'with shrinkage and numerous demolition programs, vacant spaces became very important' (ibid, p. 8), considering demographic decline that at least balances the proportion of open space for the remaining population. However, in these cases the emptiness of sites all over the city can also trigger a spiral of population decline associated with crime and squatting, which after a while spreads across the whole city. If this process extends beyond what is possible to be controlled – somehow what happened in Detroit or in some 'ghost towns' in China – the emptiness takes the scale of the entire urban area, and so, the whole city becomes interstitial (Batty, 2016; Skidmore, 2014; Xie et al., 2018).

As part of the same phenomenon, functional obsolescence also contributes to the emergence of interstitial lands and their deterioration. This is visible in emerging economies where financial asymmetries between public and private sectors restrain planning interventions on undeveloped lands. Additionally, planning regulations are often set to promote expansions instead of urban regenerations considering high costs associated to revamping previous infrastructure or pollution (Silva, 2017). This situation mostly refers to areas with diminished functionality or simply in an economic stalemate (Frantál et al., 2015). In terms of density and locations, these spaces are perceived as unpleasant considering circulation of trucks and the presence of large parking areas. In this category are brownfields and landfills or other obsolete spaces that require large investments to be revamped (Mathey et al., 2015). Over the years, these interstitial spaces are socially accepted and naturalised as part of the suburban landscape. According to Ige and Atanda (2013), urban growth in developing countries is not always accompanied by a balanced proportion of services in interstitial spaces. So, derelicts interstices become informally occupied or used as landfills (Ige & Atanda, 2013). Yet, brownfield interstitial spaces in the same inner-city locations may be subject to planning 'blight' despite their latent economic value. Here any growth machine logic is one in which local and national state expenditures rather than purely private interests may come to the fore (Phelps & Wood, 2011). The rent gap associated with derelict or unused land and properties may not close without planning intervention (to clean up or provide access to sites) and even then, may necessitate an urban politics and planning permissive of informal, temporary, or interim uses where decline is prolonged (Dubeaux & Sabot, 2018).

Some lands without restrictions to be developed can nevertheless remain undeveloped considering windfall gain. This is exacerbated in places where there is not taxation on empty lands or impact fees for regressive investment. Some of these interstitial lands are further commodified by plans and urban design projects that contemplate further profits on the potential future development (Stanley, 2014). There is enough evidence of land banking strategies as leitmotif for the creation of interstitial spaces (Van der Krabben & Jacobs, 2013). The practice of buying undeveloped lands for investment – including value capture, approval of planning, or subdivisions to then sell it at a profit – is strongly linked to urban containments that leave lands undeveloped. These mechanisms create rings of expensive undeveloped lands near the city (Han et al., 2019). Public institutions also speculate with interstitial lands gathered as reserves for future facilities, social housing developments, or infrastructure. Their condition as undeveloped is assumed by local communities as opportunities for community projects, such as gardening, city farming, parking areas, residential yards, or sports fields (O'Callaghan & Lawton, 2016), and rejecting informal occupations (Kremer & Hamstead, 2015).

Other interstices emerge from 'leapfrogging' processes of urbanisation as developing lands in outer locations leave undeveloped spaces between old and new urbanisations (Holcombe, 1999). This is driven by housing affordability policies and the provision of open space near social housing developments. So, residents accept longer commutes in exchange for lower-priced houses, and interstitial spaces become 'the commuting space' (Gallent & Shaw, 2007).

4.3.4 Policy-based determinants

As part of normative planning, undeveloped areas appear in master plans for future growth or in the form of buffer zones, industrial, residential, institutional growth, or future public spaces. When the city grows in different directions, these areas, however, remain undeveloped. It is not uncommon to see development initiatives where 'the plan remained on paper. Ironically, market forces then produced an amorphous spread city' (Hall, 2002, p. 11). This is the case of market-driven planning systems where urbanisation takes place in areas where construction norms are more flexible; systems where 'the private sector enjoys asymmetrical power in society when compared with interests concerned with the reproduction of labour' (Phelps, 2012, p. 26). This (supposed) failure of normative planning to control urban development while planning interstitial spaces relates to the fact that 'traditional planning is about maintaining the existing social order rather than challenging and transforming it' (Albrechts, 2015, p. 510). These critics also highlight a tradition of pragmatic negotiations around contingencies within a policy context shaped by 'projects' (rather than policies) and political disparities among public authorities (Faludi, 2000; Hersperger et al., 2018). These emergences of unplanned/planned interstitial spaces reflect a *status quo* in framing normative approaches to urban realities that are intrinsically diverse, dynamic, and specific.

In the outer suburbs and peri-urban areas, more land-use change occurs mainly because development pressures are more intense. Here, open spaces disappear because of the conversion of remaining raw land into developed land for the first time. While nature is often uprooted with vegetation removed or new species imported, the green spaces that are left or which were created in the development process will also retain a value in terms of their potential biodiversity and their ecosystem services (Douglas, 2008). Some regulatory systems have been relatively successful in producing interstitial spaces as part of the urban system, such as the 'Finger Plan' of Copenhagen, for instance. Since its origins, the plan was oriented to safeguard important fragments of the countryside while promoting urban growth (Caspersen et al., 2006; Gravsholt Busk et al., 2006). This balance is morphologically defined by the creation of densified zones along transport corridors ('fingers') interspersed with protected open tracts ('wedges') that penetrate the city closer to its main core. These protected tracts are composed of natural landscapes, ecological corridors, farmlands, and temporary low-density activities related to leisure and recreation (Vejre et al., 2007). Another example is Oregon's comprehensive 'Urban Growth Boundary'

(UGB) approach. The policy has been recognised as the outcome of specific forms of environmental governance in Oregon, and subsequently Portland (Lang, 2002; Lang & Hornburg, 1997). Between the 1950s and the 1970s, political authorities echoed the claims of organised campaigners – including coalitions of farmers and environmentalists – to set multiple goals governing urbanisation and the preservation of interstitial productive lands (Huber & Currie, 2007).

Planned interstitial spaces are also protected green corridors, buffers of security, regional parks, and natural reserves that are positive in terms of socio-environmental and economic benefits. According to Talen (2010), researchers have drawn connections between sprawl and regulations that derive in plans and zoning to protect valuable undeveloped land. Planners have also strategically identified interstitial areas as 'buffers' between incompatible functions such as industries and residences (Zhang et al., 2013). These buffers emerge as 'cushioning space' of differing functions: 'there are many examples of this phenomenon, such as residential zones adjacent to eight-lane freeways, and public amenities surrounded by low-density, single-family zoning. In most cases, a more appropriate spatial pattern would put open space or more resilient uses adjacent to freeways, and higher-intensity land uses adjacent to public amenities' (Talen, 2010, p. 179). Thus, open areas are also 'planned' as interstitial lands for absorbing impacts and protecting new development; the emptiness of these buffers is defined in their roles as urban containment mechanisms. In the South African context, interstitial spaces were also implemented to socially segregate cities, so, they 'formed part of the apartheid city planning strategies to separate different racial groups. As a result, a series of buffer zones were created between neighbourhoods' (Hugo & Du Plessis, 2020, p. 593).

A less evident category of planned interstitial spaces is the administrative boundaries between different districts. Many boundary areas describe ambiguous governance arrangements. The location of services, jurisprudence, land taxation, and land regulations are a matter of debate in boundary areas considering their political and functional implications (Beall et al., 2019). The location of a school or a hospital in a boundary area, for instance, result in positive externalities being borne largely by the populations of neighbouring areas. This operates at local, central, regional, national, and even international levels of jurisprudence. These are the case of inter-municipal lands, for instance, in which boundary investments are relegated to central levels or for metropolitan facilities (Becker, 1966; Delgado et al., 2008). At international levels, cross-border urbanised regions leave boundary lands for border controls and administrative offices. In a similar vein, regions with military conflicts define different types of boundaries or 'ceasefires zones' without any installation to establish interstitial lands often occupied by UN staff, NGOs, or international organisations (Newman, 1997).

4.3.5 *Infrastructural determinants*

Infrastructure of networks – such as that pumping stations, telephone exchanges, electricity power plants, railway services, and motorways – are 'often closed and recycled, as cities sourced their power and water resources from further afield (...) the huge technological networks of ducts, pipes, conduits and wires were themselves relegated to the urban background' (Graham, 2000, p. 183). Infrastructural lands can also take the form of security buffers, train stations, airports, or other large architectonic artefacts with restricted accessibility (Silva, 2017).

These infrastructural spaces have become increasingly visible due to their interstitial condition, somehow, a by-product of 'splintering urbanism' (Graham & Marvin, 2001). The presence of specific infrastructures within – or connected to – the urban fabric of cities and regions determines their interstitial character as spaces of mediation and transition of economic, social, and cultural assets, or spaces that would increasingly become built-up: 'with the ubiquity of highways, automobiles, and trucks, the interstitial areas between the older radial prongs of growth tended to fill in, generally, however, at somewhat lower densities than the older radial prongs' (Mayer, 1969, p. 15). These infrastructures can emerge as different planning responses but become interstitial when surrounded by suburban developments. They can connect different interstitial spaces too, and also circumscribe a space and become the boundary objects of an interstitial space. These are the cases of roundabouts, ring-roads, motorways, and railways or other more spatially complex spaces – such as conurbations – that become part of the sprawling urban development.

Lands with restricted access also produce interstitial spaces. These areas emerge when the urban expansion reaches interstitial hubs in the form of outer installations such as military facilities, chemical and/or power (nuclear) plants, mining sites, high-tech enclaves, upper-class gated communities, and heavy industries. Generally, these spaces are planned and managed by central authorities or corporative bodies that operate at national and international levels such as defence, public works, and industries. The emergence of these spaces in the fabric of cities and regions raises questions about their accessibility as well as the scale of their interstitial condition as they extend to non-urban geographies. Military facilities, for instance, are closed to public access, but their geographical dominium overpass immediate surroundings (Bagaeen, 2006). It is noticed that military facilities located in suburban areas are ambivalent infrastructures as they reinforce social values of national security while functionally inaccessible or incompatible with surrounding land-uses. Their fenced character linked to military operations and functions is also perceived as a sort of social denial (Warf, 1997). A similar situation is seen with airports and other interstitial spaces that operate as communication nodes. Mining sites, for instance, describe environmental alterations with impacts at global scales (Arboleda, 2016).

The fact that these types of interstices are aimed to inform studies of surrounding cities is analytically reductive, as it obscures the significance of the interstices in the intertwined global networks and supply chains that fuel the sprawling developments of cities elsewhere. The reconversion of these interstitial lands into more integrated places to the urban fabric talks about the political perils around long-term political cycles (Ponzini & Vani, 2014).

4.4 The significance of interstitial spaces

The significance of interstitial spaces has often been discussed in a fragmentary way. Sometimes, they are incidentally invoked – without providing any definition – as a sort of 'gap' within the urban fabric. Second, they appear as abandoned spaces used by marginalised groups linked to informal occupations. Third, the interstices carry a negative connotation as synonym of spatial, social, and environmental fragmentation, segregation, and environmental pollution. Fourth, the interstices are used in a positive way when describing their potential as open spaces, green infrastructure, and elements that contribute to urban wildlife and resilience in contexts of climate change. Fifth, the interstitial spaces appear as opportunities for implementing retrofitting and infilling policies in contexts of increasing demands for housing and infrastructure. Finally, the interstices are discussed as elements of 'planetary urbanisation' while referring to the regional (and global) space between cities and regions for economic trade, global circulation of goods, and global connectivity (Phelps, 2017).

4.4.1 Incidental spatial gaps

The term 'interstitial space' has been incidentally used in the planning literature while referring to issues of sprawling suburbanisation. In this case, 'interstitial spaces' are invoked to describe the by-products of urbanisation and are claimed to be re-integrated into the urban. Mohammadi et al. (2012), for instance, while discussing the sprawling development of the city of Urmia – Iran – indicates that 'sprawl leaves behind numerous "interstices" that may be used for other functions such as agricultural land or for infilling policies, although the former appears more important that the latter' (87). Similarly, Gallent and Shaw (2007) describe the rural-urban fringes of cities and how spatial planning can manage the opportunities left by the near-urban 'interstitial landscapes' (617). So, bodies of water, protected wetlands, forests, parks, slopes, open tracts, undeveloped lands, cultivated zones, sterile lands, freeways, green/open corridors, public reservations, and facilities are some of these interstitial spaces that influence the physical discontinuity of sprawling areas (Galster et.al., 2001; Zhang et al., 2013).

4.4.2 The idle lands

In the context of cities, interstitial spaces also relate to negative impacts of undeveloped areas and marginalised open tracts. Their sense of derelict and unoccupied lands is described as factors of general urban deterioration as 'the presence of abandoned land is immediately associated with increased criminal or illicit activity, physical signs of decay; and dumping of waste and conceptually linked to suburban sprawl' (Foo et al., 2014, p. 158). Ardiwijaya et al. (2014) refer to the interstices as by-products of urban sprawl in the form of abandoned 'idle lands' (p. 208). Hugo and Du Plessis (2020) indicate that 'urban interstitial spaces are often defined as anti-space, latent space, or dormant spaces' (p. 592); inefficient lands that need to be better managed, integrated, or urbanised at some point. The condition of interstices as 'vacant lands' is varied and 'includes not only publicly owned and privately owned unused or abandoned land or land that once had structures on it, but also the land that supports structures that have been abandoned, derelict, boarded up, partially destroyed, or razed' (Pagano & Bowman, 2000, p. 2). This understanding of interstitial spaces as idle lands directly connects to Foucault's heterotopic character of interstitiality.

4.4.3 The interstices as socio-cultural spaces

The literature also indicates that interstitial spaces can be integrated into the urban fabric as public spaces that support social integration. In some cities, interstitial spaces serve as platforms for activities linked to tourism, education, and research (Sieverts, 2011; van Leeuwen & Nijkamp, 2006), with clear impacts on the local economy and functional intensification. Empirical evidence suggests that these interstices support positive social and psychological processes linked to stress relief, enhancement of spiritual sensitivity, and other immaterial benefits (Banzhaf, 2007; Chiesurahico, 2004). Interstitial spaces also provide 'specific features' that make places different, a necessary condition for homogeneous suburbia resulting from the massive production of standardised houses and infrastructures (Home et al., 2010). A variety of spatial qualities are provided by interstitial spaces such as interiority, intimacy, variety of colours and light, clarity, darkness, narrowness, amplitude, depth, diversity of perspectives, textures, and others used by landscape designers to improve suburban quality (Bowler et al., 2010; Wickham et al., 2010). In a similar vein, open interstitial spaces influence the sense of safety as they provide visual dominance, orientation, and mobility (Stamps & Smith, 2002).

In a sociological dimension, Brighenti (2013) describes the interstices as a network of somehow marginal spaces where reactions against official institutions or societal anomalies can flourish. The interstices are not only spatial but also social entities – gaps within the socio-political establishment – where those excluded from society can live. Different marginalised groups can use the interstices as boundaries between themselves or between them and the formal society. The interstices are 'disorganised environments' that do not belong to the official private nor public space and serve as shelters or trenches for social struggles and vindications (Brighenti, 2013). It is important to notice that 'the ways in which informal urbanism flourishes in the spatial interstices of the city and produces urban phenomena with a potent impact on the streetscape and urban image' (Dovey, 2012, p. 352) highlights the multifaceted nature of the interstices and their contrast with the formal city. Functions that include trading, parking, hawking, begging, and advertising are mixed with planned functions determined by formal plans and designs. The interstices are sometimes informally invaded places that describe a dialectic and incongruent image of differing rationalities in the production and use of the urban space. In a similar vein, Shaw and Hudson (2009) identify 'interstitial spaces' as scenarios for artistic expressions against formal institutional controls. They highlight the creative ways in which the interstitial spaces are occupied and how they challenge orthodoxies around 'place-making' and social order. The interstitial spaces positively interfere with the controlled urban space while highlighting the social dereliction of marginalised groups. The interstices are indistinctly used by either marginal or more structured groups but in any case, have the potential to become scenarios for creativity and alternative expressions of social organisation (ibid).

On a wider scale, Tonnelat (2008) defines the urban interstices as 'zones of transition' in which immigrants learn about the local culture and obtain clues for socio-cultural adaptation to the American society before moving into more permanent residences. This situation takes place in various residual spaces between industrial facilities, roads, canals, and the poor tenements occupied by workers. This view is contrasted with the European context, in which residual interstitial spaces are seen as opportunities – often in the eye of professional design and landscape practices – to encourage community empowerment and formalisation of the space (Tonnelat, 2008).

4.4.4 Interstitial spaces as open and green infrastructure

From an environmental perspective, interstitial spaces become structural elements that may play a key role in strengthening metropolitan resilience not only from an ecological perspective but also from its multi-functional potential (Santos, 2017). Empirical studies demonstrate that interstitial spaces are important for the reduction of natural disasters, above all in highly densified areas without natural surfaces for draining storm water or for facing storm and flooding events. A soil taxonomy is indeed suggested to document the specific capacity of interstices to retain infiltration, improve absorption, and drain large amounts of water (Barkasi et al., 2012). It is argued that 'identifying and using these unused and underutilised interstitial space networks present the potential to insert small scale climate change adaptation and mitigation strategies within the city, while retaining its overall functionality (Hugo and Plessis, 2020, p. 593).

Environmental studies suggest that interstitial spaces can support the protection of natural ecosystems – including wildlife, natural landmarks, and open spaces (La Greca et al., 2011; Lafortezza, 2013; Sandström, 2002) and create interconnected networks of green spaces with benefits for both human wellbeing and nature (Lafortezza et al., 2013). This green infrastructure includes natural but also intervened spaces: 'farmlands and woods and shrubs play an important role in controlling evapotranspiring processes and in mitigating urban pollution inside highly urbanized settlements' (La Rosa & Privitera, 2013, p. 96). These interstices help to '... preserve biodiversity, sequester CO_2, produce O_2, reduce air pollution and noise, regulate microclimates, reduce the heat island effect' (La Rosa & Privitera, 2013, pp. 94–95). Suburban interstitial spaces offer possibilities for green infrastructure in an 'urbanised countryside' (van Leeuwen & Nijkamp, 2006) and contexts of 'suburban rurality' (Silva, 2020). These benefits also apply to inner suburban areas considering that 'high residential densities with large interstitial green spaces and small backyards will minimize the overall ecological impact of that city' (Lin & Fuller, 2013, p. 1164).

From a human ecology perspective, Gandy (2011) describes the 'interstitial places' of Berlin as wild, unregulated, and unexplored in terms of ecological potentials. The interstices of Berlin have an important aesthetic and scientific significance considering their micro-climates and flora and fauna associated with natural heritage. They provide valuable information about trees, local grass, stones, meadows, and other knowledge transferable among citizens. Gandy (2011) stresses the values of the interstices as they 'reveal a city within a city that is not stage-managed for tourism or consumption but open to multiple alternatives' (p. 152). The interstitial spaces configure a network of unregulated spaces where 'ecological and socio-cultural diversity can flourish' (ibid, p. 152). In a sensitive description, Gandy highlights specific details – including the wind, the grass, humidity, insects, and flowers – that create a special urban atmosphere full of new aesthetic elements. In a similar vein, Jorgensen and Tylecote (2007) assume that 'urban interstices' exist in cities as spaces for wildlife. So, woodlands, abandoned allotments, river corridors, brownfield sites, and others emerge as places for spontaneous growth of vegetation in contrast to those planned spaces with nature 'under control'. The interstices facilitate direct contact of urban dwellers with wild nature while providing alternative insights for landscape planning and urban design. This 'interstitial wilderness' reinforces the character of interstices as spaces of multiple human ecologies (Jorgensen & Tylecote, 2007). The values associated with the environmental meanings of interstitial spaces indicate that – if codified in planning – their impacts 'go beyond the environmental pollution and climate change challenges' (La Greca et al., 2011, p. 2193); interstitial spaces can increase urban quality by the creation of more socially friendly and visually pleasant suburban areas.

4.4.5 Opportunities for urbanisation and preservation

Interstices are in a tension between being preserved as open tracts or transformed into built-up space. Despite their role in recreation and conservation, the interstices are seen as options for future development, especially by developers and public entrepreneurs. Consequently, an essential continuous conflict exists between development and the conservation of interstices' (Maruani & Amit-Cohen, 2007). For some, the market value of developed land (and its speculative value affected by expectations of future development) is much higher than the value of land conserved as open space (Nelson, 1999). This transitional condition of interstitial spaces is described as a natural process of optimisation of land-use as discontinuous landscapes describe 'a haphazard patchwork, widely spread apart and seeming to consume far more land than contiguous developments. Unless preserved or unbuildable, the remaining open tracts are usually filled with new developments as time progresses' (Gillham, 2002, pp. 4–5). Oueslati et al. (2015) argue that planning policies must reduce 'the outward growth of cities and therefore encourage in-fill development in the interstices between fragments' (p. 1610).

However, the interstices as open tracts are also described as 'the actual creative field, which must preserve and restore the identity, the unique character of the *Zwischenstadt*' (Sieverts, 2003, pp. 121–122). Santos (2017) defines the interstices as 'discrete landscapes'; those that emerge besides of (and by contrast with) those 'areas where intensive land use coexists with the latent and slowly changing fabrics of rural land, often adjacent to sensitive and protected areas of ecological value' (p. 7). Less densified areas besides highly densified ones can also be considered as 'discrete interstitial spaces' – above all in heavily urbanised regions – considering that urban intensification, retrofitting, and regeneration cannot reach every corner of the sprawling suburbia. Somehow, this is another effort to define the interstices as context-dependent entities that would influence planning and design agendas of preservation and change. In that sense, Sousa-Matos (2009) understands that interstitial spaces emerge as by-products of the uncontrolled urban expansion that must be reclaimed for new developments, functions, and activities while capitalising on their current functional and spatial potentials (Sousa-Matos, 2009).

4.4.6 Bridge spaces between regional and planetary interchange

At a wider scale, the interstitial spaces are the regional (and global) scope between different settlements and regions where regional and global interchanges of different sorts occur. These scales of interstitiality have been highlighted by studies from economic geography and political ecology, indicating that 'in many cases the presented types (of interstitial spaces) imply difference of scale, from local (urban), through metropolitan (agricultural and rural) and regional (countryside), and up to a national scale (wilderness)' (Maruani & Amit-Cohen, 2007, p. 6). It has been observed that these 'strange places exist outside the city's effective circuits and productive structures' (de Sola-Morales 1995, p. 120), and from an economic point of view they represent places, 'where the city is no longer' (ibid, p. 120). However, the activities and functions of these regional interstices influence the urban character of cities – not only those that surround the regional space – but also cities elsewhere. Somehow, this is the 'interstitial geography' that emerges as an extension of the city's economic and cultural dynamic over the countryside. These spaces operate as 'bridges'; spaces that 'enjoy a highly strategic and centralised position, even though they boast few links' (Pflieger & Rozenblat, 2010, p. 2727), while composed of public transport systems, convergence, and intersection of different social groups, economic enclaves, and other relational elements.

4.5 Conclusions

Interstitial spaces are not random outcomes of less controlled processes in planning. Even under conditions of urban growth, the competing economic, social and environmental values adhering to interstitial spaces ensure that some spaces – including the largest and most desirable to the private sector – can be pending for prodigious periods. This 'pending' condition shapes them as interstitial although transitioning to become built-up spaces of some sort or another. Although this transition is more evident in the suburban realm of infilling practices, in the regional interstitial space between cities these transitions take the form of interlinked networks of enclaves and infrastructures. The different temporalities of the interstices at suburban, regional, and global realms talk about the importance of the scales at which the interstitial spaces can manifest. As such, the scales also define the relational character of the interstices as entities of interaction between different forces and actors placed within and beyond the geography of cities and their planning reasoning for the built-up space.

References

Albrechts, L. (2015). Ingredients for a more radical strategic spatial planning. *Environment and Planning B: Planning and Design, 42*(3), 510–525.

Angelo, H., & Wachsmuth, D. (2015). Urbanizing urban political ecology: A critique of methodological cityism. *International Journal of Urban and Regional Research, 39*(1), 16–27.

Arboleda, M. (2016). In the nature of the non-city: Expanded infrastructural networks and the political ecology of planetary urbanisation. *Antipode, 48*(2), 233–251.

Ardiwijaya, V. S., Soemardi, T. P., Suganda, E., & Temenggung, Y. A. (2014). Bandung Urban sprawl and idle land: Spatial environmental perspectives. *APCBEE procedia, 10*, 208–213.

Bagaeen, S. (2006). Redeveloping former military sites: Competitiveness, urban sustainability and public participation. *Cities, 23*(5), 339–352.

Banzhaf, H. S. (2007). 'Public benefits of undeveloped lands on urban outskirts: Non-market valuation studies and their role in land use plans'. Andrew Young School of Policy Studies Research Paper Series, (07–28).

Barkasi, A., Dadio, S., Losco, R., & Shuster, W. (2012). Urban soils and vacant land as stormwater resources. *World Environmental and Water Resources Congress 2012*. American Society of Civil Engineers. 569–579.

Batty, M. (2016). Empty buildings, shrinking cities and ghost towns. *Environment and Planning B: Planning and Design, 43*, 3–6.

Beall, J., Cherenet, Z., Cirolia, L., & da Cruz, N. F. (2019). Understanding infrastructure interfaces: Common ground for interdisciplinary urban research? *Journal of the British Academy, 7*(s2), 11–43.

Becker, D. M. (1966) Municipal Boundaries and Zoning: Controlling Regional Land Development. Wash. ULQ, 1.

Boland, P., Bronte, J., & Muir, J. (2017). On the waterfront: Neoliberal urbanism and the politics of public benefit. *Cities, 61*, 117–127.

Bowler, D., Buyung-Ali, L., Knight, T., & Pullin, A. (2010). Urban greening to cool Towns and cities: A systemic review of the empirical evidence. *Landscape and Urban Planning, 97*, 147–155.

Brighenti, A. M. (2013). *Urban interstices: The aesthetics and the politics of the in-between.* Ashgate.

Bruegmann, R. (2005). *Sprawl: A compact history.* University of Chicago Press.

Burchfield, M., Overman, H. G., Puga, D., & Turner, M. A. (2006). Causes of sprawl: A portrait from space. *The Quarterly Journal of Economics, 121*(2), 587–633.

Caspersen, O. H., Konijnendijk, C. C., & Olafsson, A. S. (2006). Green space planning and land use: An assessment of urban regional and green structure planning in greater Copenhagen. *Geografisk Tidsskrift-Danish Journal of Geography, 106*(2), 7–20.

Chiesurahico, A. (2004). The role of urban parks for the sustainable city. *Landscape and Urban Planning, 68*(1), 129–138.

Choy, D. L., & Buxton, M. (2013). Farming the city fringe: Dilemmas for peri-urban planning. In Q. Farmar-Bowers, V. Higgins, & J. Millar (Eds.), *Food security in Australia.* Springer.

Clawson, M. (1962). Urban sprawl and speculation in suburban land. *Land Economics, 38*(2), 99–111.

de Sola-Morales, I. (1995). Terrain vague. In C. Davidson (Ed.), *Anyplace* (pp. 118–123). MIT Press.

Del Castillo, J., & Castro, M. (2015). La estrategia de integración de espacios abiertos y patrimonio en el plan metropolitano de desarrollo urbano Lima-callao 2035. *Devenir-Revista de estudios sobre patrimonio edificado, 2*(4), 27–44.

Del Castillo, J. M., & Sopla, P. (2018). Paisajes prehispánicos intersticiales: Naturaleza Urbana y patrimonio arqueoastronómico en Ñaña, Lima [prehispanic interstitial landscapes: Urban nature and archeo-astronomic patrimonia]. *Devenir-Revista de estudios sobre patrimonio edificado*, *5*(10), 153–174.

Delgado, O. B., Mendoza, M., Granados, E. L., & Geneletti, D. (2008). Analysis of land suitability for the siting of inter-municipal landfills in the Cuitzeo Lake basin, Mexico. *Waste Management*, *28*(7), 1137–1146.

Douglas, I. (2008). Environmental change in peri-urban areas and human and ecosystem health. *Geography Compass*, *2*(4), 1095–1137.

Douglas, I. (2016). Urban ecology. In D. Richardson, N. Castree, M. F. Goodchild, A. Kobayashi, W. Liu, R. A. Marston, and K. Falconer Al-Hindi (Eds.), *International Encyclopedia of Geography: People, the Earth, Environment and Technology*, (pp, 7319–7330). John Wiley & Sons Inc.

Dovey, K. (2012). Informal urbanism and complex adaptive assemblage. *International Development Planning Review*, *34*(4), 349–367.

Dubeaux, S., & Sabot, E. C. (2018). Maximizing the potential of vacant spaces within shrinking cities, a German approach. *Cities*, *75*, 6–11.

Faludi, A. (2000). The performance of spatial planning. *Planning Practice and Research*, *15*(4), 299–318.

Faludi, A. (2005) The Netherlands: A country with a soft spot for planning. In B. Sanyal (Ed.), *Comparative planning cultures* (pp. 309-332). Routledge.

Foo, K., Martin, D., Wool, C., & Polsky, C. (2014). The production of urban vacant land: Relational placemaking in Boston, MA neighborhoods. *Cities*, *35*, 156–163.

Frantál, B., Greer-Wootten, B., Klusáček, P., Krejčí, T., Kunc, J., & Martinát, S. (2015). Exploring spatial patterns of urban brownfields regeneration: The case of Brno, Czech Republic. *Cities*, *44*, 9–18.

Gallent, N., & Shaw, D. (2007). Spatial planning, area action plans and the rural-urban fringe. *Journal of Environmental Planning and Management*, *50*(5), 617–638.

Galster, G., Hanson, R., Ratcliffe, M., Wolman, H., Coleman, & Freihage, J. (2001). Wrestling sprawl to the ground: Defining and measuring an elusive concept. *Housing Policy Debate*, *12*(4), 681–717.

Gandy, M. (2011). Interstitial landscapes: Reflections on a Berlin corner. In M. Gandy (Ed.), *Urban constellations* (pp. 149–152). Jovis.

Gandy, M. (2015). From urban ecology to ecological urbanism: An ambiguous trajectory. *Area*, *47*(2), 150–154.

Gillham, O. (2002). *The limitless city: A primer on the urban sprawl debate*. Island Press.

Godfrey-Smith, W. (1979). The value of wilderness. *Environmental Ethics*, *1*(4), 309-319.

Graham, S. (2000). Constructing premium network spaces: Reflections on infrastructure, networks and contemporary urban development. *International Journal of Urban and Regional Research*, *24*(1), 183–200.

Graham, S., & Marvin, S. (2001). *Splintering urbanism: Networked infrastructures, technological mobilities and the urban condition*. Psychology Press.

Gravsholt Busk, A., Kristensen, S., Praestholm, S., Reenberg, A., & Primdahl, J. (2006). Land system changes in the context of urbanization: Examples from the peri-urban area of greater Copenhagen. *Danish Journal of Geography*, *106*(2), 21–34.

Green, O., Garmestani, A., Albro, S., Ban, N., Berland, A., Burkman, C., & Shuster, W. D. (2016). Adaptive governance to promote ecosystem services in urban green spaces. *Urban Ecosystems*, *19*(1), 77–93.

Hall, P. (2002). *The Buchanan report: 40 years on. Transport, 157*, 7–14.

Han, M. Y., Chen, G. Q., & Dunford, M. (2019). Land use balance for urban economy: A multi-scale and multi-type perspective. *Land Use Policy, 83*, 323–333.

Harvey, T., & Works, M. A. (2002). Urban sprawl and rural landscapes: Perceptions of landscape as amenity in Portland, *Oregon. Local Environment, 7*(4), 381–396.

Hersperger, A. M., Oliveira, E., Pagliarin, S., Palka, G., Verburg, P., Bolliger, J., & Grădinaru, S. (2018). Urban land-use change: The role of strategic spatial planning. *Global Environmental Change, 51*, 32–42.

Hess, G., Daley, S. S., Dennison, B. K., Lubkin, S. R., McGuinn, R. P., Morin, V. Z., & Wrege, B. M. (2001). Just what is sprawl, anyway. *Carolina Planning, 26*(2), 11–26.

Heynen, N., Kaika, M., & Swyngedouw, E. (2006). Urban political ecology. In K. Heynen, & E. Swyngedouw (Eds.), *In the nature of cities: Urban political ecology and the politics of urban metabolism* (pp. 1–20). Routledge.

Hill, I. T. (2013). *The ancient city of Athens.* Harvard University Press.

Holcombe, R. (1999). In defense of urban sprawl. Enhancing the quality of Life. *Urban Sprawl: Pro and Con.* PERC REPORTS, *17*(1), 1–20.

Home, R., Bauer, N., & Hunziker, M. (2010). Cultural and biological determinants in the evaluation of urban green spaces. *Environment and Behavior, 42*(4), 494–523.

Hong, W., Yang, C., Chen, L., Zhang, F., Shen, S., & Guo, R. (2017). Ecological control line: A decade of exploration and an innovative path of ecological land management for megacities in China. *Journal of Environmental Management, 191*, 116–125.

Huber, M. T., & Currie, T. M. (2007). The urbanization of an idea: Imagining nature through urban growth boundary policy in Portland. *Oregon. Urban Geography, 28*(8), 705–731.

Hugo, J., & Du Plessis, C. (2020). A quantitative analysis of interstitial spaces to improve climate change resilience in Southern African cities. *Climate and Development, 12*(7), 591–599.

Ige, J. O., & Atanda, T. A. (2013). Urban vacant land and spatial chaos in Ogbomoso North local government, Oyo State, Nigeria. *Global Journal of Human Social Science & Environmental Science & Disaster Management, 13*(2), 28–36.

Jorgensen, A., & Tylecote, M. (2007). Ambivalent landscapes – wilderness in the urban interstices. *Landscape Research, 32*(4), 443–462.

Kim, G., Miller, P. A., & Nowak, D. J. (2015). Assessing urban vacant land ecosystem services: Urban vacant land as green infrastructure in the City of Roanoke, Virginia. *Urban Forestry & Urban Greening, 14*(3), 519–526.

Koolhaas, R. (2021). *Countryside. A report. The countryside in your pocket!* Taschen.

Kremer, P., & Hamstead, Z. (2015). Transformation of urban vacant lots for the common good: An introduction to the special issue. *Cities and the Environment* (CATE), *8*(2), 1–6.

La Greca, P. L., Rosa, D., Martinico, F., & Privitera, R. (2011). Agricultural and green infrastructures: The role of non-urbanized areas for eco-sustainable planning in metropolitan region. *Environmental Pollution, 159*(01), 2193–2202.

La Rosa, D., & Privitera, R. (2013). Characterization of non-urbanized areas for land-use planning of agricultural and green infrastructure in urban contexts. *Landscape and Urban Planning, 109*(01), 94–106.

Lafortezza, R., Davies, C., Sanese, G., & Konijnendijk, C. C. (2013) Green infrastructure as a tool to support spatial planning in European urban regions. *iForest – Biogeosciences and Forestry*, 6(3), 102–108.

Lang, R. E. (2002). Does Portland's urban growth boundary raise house prices? *Housing Policy Debate*, *13*(1), 1–5.

Lang, R., & Hornburg, S. P. (1997). Planning Portland style: Pitfalls and possibilities. *Housing Policy Debate*, *8*(1), 1–10.

Li, Q., Yuichi, F., & Morris, M. (2014). Study on the buffer zone of a cultural heritage site in an urban area: The case of Shenyang imperial palace in China. *WIT Transactions on Ecology and the Environment*, *191*, 1115–1123.

Lin, B. B., & Fuller, R. A. (2013). Sharing or sparing? How should we grow the world's cities? *Journal of Applied Ecology*, *50*(5), 1161–1168.

MacLeod, G. (2011). Urban politics reconsidered: Growth machine to post-democratic city? *Urban Studies*, *48*(12), 2629–2660.

Maruani, T., & Amit-Cohen, I. (2007). Open space planning models: A review of approaches and methods. *Landscape and Urban Planning*, *81*(1), 1–13.

Mathey, J., Rößler, S., Banse, J., Lehmann, I., & Bräuer, A. (2015). Brownfields as an element of green infrastructure for implementing ecosystem services into urban areas. *Journal of Urban Planning and Development*, *141*(3), A4015001.

Mayer, H. M. (1969). Cities and urban geography. *Journal of Geography*, *68*(1), 6–19.

Ministry of Infrastructure and the Environment (2012). *35 Icons of Dutch Spatial Planning*. At: https://www.government.nl/documents/leaflets/2012/12/19/icons-of-dutch-spatial-planning (accessed in May 2021).

Mohammadi, J., Asghar, A., & Mobaraki, O. (2012). Urban sprawl pattern and effective factors on them: The case of Urmia City, Iran. *Journal of Urban and Regional Analysis*, *04*(1), 77–89.

Nelson, A. (1999). Comparing States with and without growth management analysis based on indicators with policy implications. *Land Use Policy*, *16*, 121–127.

Newman, D. (1997). Creating the fences of territorial separation: The discourses of Israeli-Palestinian conflict Resolution. *Geopolitics*, *2*(2), 1–35.

Northam, R. (1971). Vacant land in the American City. *Lands Economics*, *47*(4), 345–355.

O'Callaghan, C., & Lawton, P. (2016). Temporary solutions? Vacant space policy and strategies for re-use in Dublin. *Irish Geography*, *48*(1), 69–87.

Oueslati, W., Alvanides, S., & Garrod, G. (2015). Determinants of urban sprawl in European cities. *Urban Studies*, *52*(9), 1594–1614.

Pagano, M., & Bowman, A. (2000). *Vacant land in cities: An urban Resource* (pp. 1–9). The Brookings Institution – Survey series. Center on Urban & Metropolitan Policy.

Pflieger, G., & Rozenblat, C. (2010). Introduction. Urban networks and network Theory: The City as The connector of multiple networks. *Urban Studies*, *47*(13), 2723–2735.

Phelps, N. (2012). *An anatomy of sprawl. Planning and politics in Britain*. Routledge.

Phelps, N. A. (2015). *Sequel to suburbia: Glimpses of America's post-suburban future*. MIT Press.

Phelps, N. A. (2017). *Interplaces: An economic geography of the inter-urban and international economies*. Oxford University Press.

Phelps, N. A., & Silva, C. (2017). Mind the gaps! A research agenda for urban interstices. *Urban Studies*, *55*(6), 1203–1222.

Phelps, N., & Wood, A. (2011). The new post-suburban politics? *Urban Studies*, *48*(12), 2591–2610.

Pierce, J., & Martin, D. G. (2015). Placing Lefebvre. *Antipode*, *47*(5), 1279–1299.

Ponzini, D., & Vani, M. (2014). Planning for military real estate conversion: Collaborative practices and urban redevelopment projects in two Italian cities. *Urban Research & Practice*, 7(1), 56–73.

Roche, M., & Argent, N. (2015). The fall and rise of agricultural productivism? An antipodean viewpoint. *Progress in Human Geography*, 39(5), 621–635.

Rutherford, J. (2008). Unbundling Stockholm: The networks, planning and social welfare nexus beyond the unitary city. Geoforum, 39(6), 1871–1883.

Said, M. Y., Ogutu, J. O., Kifugo, S. C., Makui, O., Reid, R. S., & de Leeuw, J. (2016). Effects of extreme land fragmentation on wildlife and livestock population abundance and distribution. *Journal for Nature Conservation*, 34, 151–164.

Sandström, U. (2002). Green infrastructure planning in urban Sweden. *Planning Practice & Research*, 17(4), 373–385.

Santos, J. R. (2017). Discrete landscapes in metropolitan Lisbon: Open space as a planning Resource in times of latency. *Planning Practice & Research*, 32(1), 4–28.

Saxena, A., & Sharma, A. (2013). Importance of relationship between built forms amidst open spaces in historical Areas. *International Journal of Engineering Research & Technology (IJERT)*, 2(2), 1–12.

Shaw, P., & Hudson, J. (2009). The qualities of informal space: (Re)appropriation within the informal, interstitial spaces of the city. In *Proceedings of the conference occupation: Negotiations with constructed space* (pp. 1–13). University of Brighton.

Sieverts, T. (2003). *Cities without cities. An interpretation of the Zwischenstadt.* Spon Press.

Sieverts, T. (2011). The in-between city as an image of society: From the impossible order towards a possible disorder in the urban landscape. In D. Young, P. Wood, & R. Keil (Ed.), *In-between infrastructure: Urban connectivity in an age of vulnerability* (pp. 19–27). Praxis (e)Press.

Silva, C. (2017). The infrastructural lands of urban sprawl: Planning potentials and political perils. *Town Planning Review*, 88(2), 233–256.

Silva, C. (2020). The rural lands of urban sprawl: Institutional changes and suburban rurality in Santiago de Chile. *Asian Geographer*, 37(2), 117–144.

Skidmore, M. (2014). Will a greenbelt help to shrink Detroit's wasteland? *Land Lines*, 26(4), 8–17.

Sousa-Matos, R. (2009). Urban landscape: Interstitial spaces. *Landscape Review*, 13(1), 61–71.

Stamps, A., & Smith, S. (2002). Environmental enclosure in urban settings. *Environment and Behavior*, 34(1), 781–794.

Stanley, B. W. (2014). Local property ownership, municipal policy, and sustainable urban development in Phoenix, AZ. *Community Development Journal*, 50(3), 510–528.

Talen, E. (2010). Zoning for and against sprawl: The case for form-based codes. *Journal of Urban Design*, 18(2), 175–200.

Tewdwr-Jones, M. (1997). Green belts or green wedges for Wales? A flexible approach to planning in the urban periphery. *Regional Studies*, 30, 72–77.

Tonnelat, S. (2008). 'Out of frame'. The (in)visible life of urban interstices – a case study in Charenton-le-Pont, Paris, France. *Ethnography*, 9(3), 291–324.

Van der Krabben, E., & Jacobs, H. M. (2013). Public land development as a strategic tool for redevelopment: Reflections on the Dutch experience. *Land Use Policy*, 30(1), 774–783.

van Leeuwen, E., & Nijkamp, P. (2006). The urban-rural nexus: A study on extended urbanization and the hinterland. *Studies in Regional Science*, *36*(2), 283–303.

Vejre, H., Primdahl, J., & Brandt, J. (2007). The Copenhagen Finger Plan. Keeping a Green Space Structure by a Simple Planning Metaphor. Europe's living landscapes. Essays on exploring our identity in the countryside. *LANDSCAPE EUROPE/ KNNV*.

Vergara, F., & Boano, C. (2020). Exploring the contradiction in the ethos of urban practitioners under neoliberalism: A case study of housing production in Chile. *Journal of Planning Education and Research*, 1–15.

Wachsmuth, D. (2012). Three ecologies: Urban metabolism and the society-nature opposition. *The Sociological Quarterly*, *53*(4), 506–523.

Wang, S., Hong, L., & Chen, X. (2012). Vulnerability analysis of interdependent infrastructure systems: A methodological framework. *Physica A: Statistical Mechanics and Its Applications*, *391*(11), 3323–3335.

Warf, B. (1997). The geopolitics/geoeconomics of military base closures in the USA. *Political Geography*, *16*(7), 541–563.

Wickham, J., Riitters, K., Wade, T., & Vogt, P. (2010). A national assessment of green infrastructure and change for the conterminous United States using morphological image processing. *Landscape and Urban Planning*, *94*, 186–195.

Xie, Y., Gong, H., Lan, H., & Zeng, S. (2018). Examining shrinking city of Detroit in the context of socio-spatial inequalities. *Landscape and Urban Planning*, *177*, 350–361.

Zhang, R., Pu, L., & Zhu, M. (2013). Impacts of transportation arteries on land use patterns in urban-rural fringe: A comparative gradient analysis of Qixia District, Nanjing City, China. *Chinese Geographical Science*, *23*(3), 378–388.

5 The scales of the interstitiality

5.1 Introduction

> … it is a repository for buildings so big they don't fit in any city.
>
> (Koolhaas, 2021, p. 272)

It has been illustrated that interstitial spaces embrace from the space between two neighbourhoods up the space between cities and regions … and beyond. Considering that interstitial spaces cannot be reduced to their morphological characteristics (Brighenti, 2013, p. xviii), it would be necessary to notice their spatial distribution and varied magnitudes. As such, the interstitial spaces range from liminal spaces between suburban buildings at the edge of cities up to colossal landscapes between regions; a reality of interstices that demands an analytical perspective that speaks to their multiple geographical scales.

The scales at which interstitial spaces manifest are closely related to the functional capacities of land and institutional frameworks that restrict development. However, the spatial landscape at which these multiple functions are located speaks about the hinterland of the interstitial space and its elasticity, as in many cases these functions extend beyond physical locality. As pointed out by Foo et al. (2013) 'The spatial heterogeneity of urban redevelopment, including the variability in which vacant areas are prioritized for growth, calls for a multi-scale analysis that examines changing functions of vacant land as they relate to local, urban, and regional dynamics' (p. 156). These different scales frequently determine interstitial spaces as soon to be revamped, definitively abandoned or protected, or perceived as a commodity, but also about their functions connected to network spaces. So, rural spaces, restriction areas, industrial and military facilities, buffers of security, infrastructural lands, ecological reservations, mining sites, and others have different sizes and are connected to different local, regional, and global hinterlands. These scales and boundaries play differing roles in planning agendas and define different levels of integration including

DOI: 10.4324/9780429320019-5

from totally closed environments until open pieces of countryside without restrictions to accessibility.

Abstracting and simplifying from the conceptual and empirical grounds that explain the emergence of interstitial spaces and their characteristics, this chapter proposes four different scales at which the interstitial spaces can manifest: (a) *the scale of proximity*, (b) *the scale of transition*, (c) *the regional scale*, and (d) *the scale of remoteness*. These four scales depict the range of sizes, transformations, and interconnection of interstitial spaces with their surroundings and wider hinterlands. This chapter focusses on these four scales as explicative of the urban condition of sprawling city-regions. The architectonic scale is not depicted as useful for understanding urbanisation, and thus, the space that lies between buildings, furniture, two trees in a street, and others is not matter of enquiry of this chapter. This scale of interstitiality has been sufficiently commented and reasonably (although succinctly) addressed by some streams of architectural research in both technical and socio-spatial terms (Gilley, 2006; Steele & Keys, 2015). The *global* scale – the one that someone can suggest beyond the proposed scale of *remoteness* – is not consigned as such as it is included into the scale of *remoteness* while often invoked to refer to the process of globalisation of cities specifically (Bocco, 2016). Despite this, the focus of this theoretical approach intends to be away from any city-centred view as explicative of the urban condition. In that sense, the scale of *remoteness* transcends the analytical scope of cities and regions and mobilises the reflection of interstitial spaces up to the global space of networks.

5.2 The scales of interstitial spaces

Ranging from liminal spaces of small size up to hinterlands of global interchange, the diverse magnitudes, morphologies, functionalities, locations, and surroundings define the role of interstitial spaces as scalar and inter-scalar entities. Inner suburban interstitial spaces described by river basins, for instance, illustrate large and narrow spaces crossing through several districts with an intricate morphology. Urban rivers describe a small distance between the two shores – a small scale – but at the same time a longer distance since the river connects several areas of the city and beyond. A narrow infrastructural interstice (a motorway or railway) or a natural feature (a river or forest) could cover a wide range of urbanised surroundings and connect several districts and have ramifications across several scales, notably in terms of governance. A large open space could exist entirely at one scale and within a single administrative jurisdiction posing few issues in terms of governance. So, the notion of 'scale' describes spatial, functional, and administrative dimensions, which implies multiple governance challenges while a single interstitial space could relate to (or integrate) more than one of these scales. This defines the significance of interstitial spaces at scalar ontologies, while 'an examination of the past, current, and potential future functions of vacant land in the context of the daily experiences for urban residents lends insight into the way that actors at multiple scales shape the uses and meanings of urban space' (Foo et al., 2013, p. 158). The scale of interstitial spaces needs to be understood in terms of the relationship between a parcel of land, space, or open tract and the wider physical and socio-economic spatial systems to which it belongs (Platt, 2004).

It has been clarified that in many cases interstitial spaces 'imply difference of scale, from local (urban), through metropolitan (agricultural and rural) and regional (countryside), and up to a national scale (wilderness)' (Maruani & Amit-Cohen, 2007, p. 6) considering that the scales of interstitial spaces also relate to the stage of expansion of a city-region. In monocentric cities, suburban interstitial spaces appear as available spaces for residential extensions or infrastructure, public spaces, or services, while in polycentric regions interstitial spaces connect to agricultural functions, industrial or protected ecosystems less controlled by the city's planning instruments. Douglas (2008) distinguishes interstitial spaces as widely distributed in fringe-belt areas, suburbs, and inner suburban locations. At the heart of large cities interstices are usually protected public and green spaces. In expansion areas changes occur more frequently as transformations are more intense. The proximity to the city or to transport infrastructure implies several pressures on land-use change. Older buildings tend to be replaced by new structures while open spaces decrease due to paving processes, increments of density, and relocation of retail and large office complexes (Kabisch & Haase, 2013). In the peri-urban area, changes affect large rural lands and ecosystem services (Douglas, 2008).

5.2.1 The scale of proximity

A first scale of interstitiality is *the scale of proximity*. This scale of interstitiality – although not strictly architectonic – is often a matter of architectural analysis as its magnitude supposes a possible extension of architectural theory into the city scale. Somehow, it has become part of the realm of urban design and local planning practice due to its closer relation to the urban core or other existing urban fragments. This scale of interstitiality is most clearly visible in monocentric cities and towns, whose most rapid urban expansion has already passed, or in economically less dynamic urban systems. The set of possibilities embraces the likes of parks, squares, protection buffers and some restriction zones, local services, and infrastructure often in inner suburban locations.

The scale of interstitial proximity describes spatial configurations where alternative (out of planned purpose) social practices can occur. These interstices also take the form of undefined open spaces between neighbourhoods in which sporting activities or other collective (although informal) functions are developed. Steele and Keys (2015) argue that 'interstitial spatial practices' that take place in suburban areas suggest further improvements on social housing policies or suburban public infrastructure. These interstitial spaces can also be the space between large facilities and buildings such as pavilions, large yards, green and blue corridors, large parking areas or unattended facilities. These interstices are key to exploring everyday neighbourhood social practices in terms of transforming their condition of non-places into Lefebvre's absolute space. This scale of interstitiality has the potential to support the emergence of functionally flexible spaces (Steele & Keys, 2015).

Insights from the literature on 'liminal landscapes' resonate with this scale of interstitiality. Liminality speaks about the in-between spaces that serve as 'thresholds' between other spaces and meaningful living experiences (Arvanitis et al., 2019), and in which different interactions occur for a specific period of time. The notion of 'liminal space' has an anthropological salience as a space of transition associated with a 'liminal experience' (the 'rite of passage'). As such, it is 'of inferential significance here insofar as it is "located" in terms of both its geography and its constitutive temporality' (Roberts, 2018, p. 40). There are several examples of interstitial spaces as 'borders' between differing functionalities that operate as liminal spaces, spaces in-between self-contained (and bounded) urban enclaves, boundaries, and borderlands within cities (Iossifova, 2013). As such, it would be reductive to conceive interstices of proximity as physical elements or abstract territorial lines; borders are about interactions between people and institutional bodies, and suppose social constraints placed by contextual and structural factors. The interstices of proximity operate as boundary objects that can be closed and rigid or can be open and permeable to facilitate the emergence of transboundary interactions and socio-spatial transitions between surrounding fragments. This also speaks about the relationality of the interstitial spaces and the infrastructures that are part of. In some cases, these infrastructures are symbolic and interstices of proximity take the form of shared public spaces of integration but also contestation and dispute between dissimilar communities; zones of contact at the edge of differentiated groups in terms of wealth and religion (Herrault & Murtagh, 2019). The liminal spaces between religious communities in Belfast – Northern Ireland – are examples of these interstitial spaces of proximity. The 'peace walls' are not mere physical elements but embedded symbols within the urban fabric that knit parks and streets. A case in point is the Alexandra Park in northeast Belfast, which is a park divided by one of the peace walls and a gate that separates Unionist/Loyalist and Nationalist/Republican communities. The gate is sometimes opened for the neighbouring communities to share the park and cross from one side to the other (Figure 5.1). This is an interstitial space of proximity that becomes liminal; a 'rite of passage' that nevertheless does not represent any easy-going sense of political integration; 'not that the search for shared space is itself a shared or coherent political project' (ibid, p. 252).

Figure 5.1 The Alexandra Park in North Belfast, Northern Ireland.

Source: Author, 2021.

The scale of proximity also describes interstitial spaces resulting from the random distribution of large social housing projects and where normative restrictions obligate developers to locate projects distanced from active farmlands. In other cases, normative restrictions on social housing require open space, so developers leave empty areas that cannot be urbanised. For Vidal (2002), these interstitial spaces are strongly influenced by the surroundings and under pressure to become built-up while reinforcing their own identity; interstitial spaces are 'a changing space in motion and a signifier of the place energy as it transfers important information about the meaning of the place. In any case, it is a mutation, and as a mutation it is a chance to create more space in the urban structure; a sort of reservoir of space for future expansions of fragments' (Vidal, 2002, pp. 162–163). This scale of proximity includes the interstitial spaces that can be integrated into the urban fabric and where a sense of 'human scale' allows recognising people among a multitude; distances where people can recognise neighbours, children, groups, elements of the natural environment, distant voices, the noise of trucks, neighbouring cars or local public transport. These interstices of proximity are not residual space, remnants, empty, outdated, deciduous, hidden, or forsaken spaces but visible, usable, and feasible to be (re)integrated into the city's fabric (Vidal, 2002). The proximity facilitates gathering different identities from surroundings, and thus, interstitial spaces become elements of individual and collective significance. As such, they are dynamic spaces that can change their own identity while taking the identity of their surroundings.

5.2.2 The scale of transition

Interstitial spaces of transition are those where undeveloped areas help define the character of contemporary urban expansion within a metropolitan system. At this second scale, the interstitial spaces present a multi-level issue for governance, implying a need for coordination across different policy sectors and districts, most commonly in a monocentric metropolitan context (Nechyba & Walsh, 2004). These interstices are the likes of large squares, sporting facilities, private and public spaces that facilitate contact with nature, metropolitan parks, military facilities, zoos, port-ship areas, industrial and research facilities, bigger restriction zones such as landfills and brownfields, large suburban farming sites, sites for the extraction of raw material, warehouses, peri-urban shopping malls, protected areas or ecosystem features such as flood plains, valleys or hills that are – or are about to be – surrounded by the sprawling urban expansion. These interstitial spaces are those located at the fringes of suburban areas up to the peri-urban space of city-regions and influence the definition of the city's urban-rural transect.

The interstitial scale of transition emerges as a good scenario for landscape management and large-scale urban design and planning projects. These interstices of transition allow the location of large housing developments, industries, and commercial facilities. Some of them are located along transport corridors that connect the interstices with wider networks and regional scales of interaction. The Belfast Harbour is an example of an interstitial space of transition with inter-scalar salience to global markets. This is an area of around 810 hectares representing 20% of Belfast City area. The harbour is considered the main maritime hub in Northern Ireland, handling 70% of Northern Ireland's seaborne trade and 20% of the maritime trade of the entire island of Ireland (Belfast Harbour Port Authority, 2021). Like all UK ports, Belfast harbour operates in a highly regulated space and is fully cognisant of the regulatory obligations linked to people's safety, environment and security. The whole harbour is an amalgamation of different interstitial hubs that include port-ship areas, industrial backyards, large warehouses, retail spaces, large parking and cargo areas, loading cranes, ship docks and terminals, large areas of containers and storage of raw material, military facilities, the George Belfast City Airport and its landing field, but also some open and real estate projects under urban regeneration schemes such as the Belfast's Titanic Quarter (Titanic Quarter Ltd., 2021). The harbour gathers several corporative and public bodies with implications on urban governance. This case speaks about the flexible and rigid character of the interstitial spaces of transitions, as while defining hard boundaries with some areas of the city, the interstices also

Figure 5.2 One of the open tracts at the Belfast Harbour, Northern Ireland, UK.

Source: Author, 2021.

have flexible boundaries connected to the global networks through maritime circulations of people and commodities (Figure 5.2).

5.2.3 The regional interstitial scale

The spaces between different settlements as part of an urbanised region define a third *regional scale* of interstitiality, which implies a correspondingly wider coordination across policy sectors and districts and a wide array of special-purpose authorities related to agriculture, industry, environment, and public works. This scale pertains to more advanced sprawling urbanised regions in which it is possible to identify satellite towns and clear conurbation zones (Morrison, 2010). As such, this scale spans various administrative levels and describes significant economic potential in the cognitive-cultural economy where 'restructuring effects in many of the interstitial spaces between large cities ... significantly redefine what it means to be rural' (Scott, 2012, p. xi).

This scale is clearly observed in polycentric regions and embraces vast areas between two or more independent cities. In this regional space, the scope of the functional linkage between work and residence extends to outer suburbs, rural lands, satellite towns, and even other urbanised regions. This is a scale where the work-residence commuting takes place outside (and beyond) cities. (Meijers & Burger, 2017). In the EU context, this space of linkage – and the cities' hinterland of influence – is the space of the FUAs, PIAs and PolyFUAs observed in the Randstad of the Netherlands or the commuting space between Manchester and Liverpool in the UK. When intensified, this space of linkage describes suburbs-to-suburbs – or inter-suburbs – connections between different cities that transgress binomial centre-periphery relations. It also refers to wider regional spaces of commuting between urbanised regions through large geographical spaces such as sea channels or large ecological reservoirs.

These regional interstices are important in terms of environmental services, economic growth, and the growth of surrounding cities beyond local jurisprudence. It has been indeed observed that – beyond the city scope – 'significant attention is being devoted to the impact of urban-economic processes on interstitial spaces lying between metropolitan areas' (Harrison & Heley, 2015, p. 1113). Zhang and Grydehøj (2020) highlight the functions of peripheral islands as interstitial hubs in the processes of urbanisation of the Zhoushan Archipelago, China. By arguing that 'newly available and repurposed transport and communication technologies facilitate inter-city, inter-region, and international connectivities that seem to challenge the status of the "city" as the critical unit in urbanism' (p. 2158). Here, the interstitial space is more an inter-urban boundary zone of exchange – an 'interface' – that articulates the economic and social dynamics of surrounding urban hubs. This is confirmed by Rickards et al. (2016) while examining the extent to which traditional understandings of 'city-region' obscure 'the interstitial spaces that lie between designated metropolitan areas but that produce a large percentage of national wealth' (p. 1539).

These regional interstitial spaces are the context of various hubs – including regional parks, large military and industrial facilities, mining sites, large venues for massive encounters, zoos, landfills and brownfields, satellite towns, exurbs, ecological areas, or geographical accidents such as flood valleys and hills – that contribute to the overall economic, environmental, and functional performance of cities and regions through modern infrastructures and network spaces. Large bodies of water – possible to be found in urban areas located in archipelagos or estuaries – also determine a type of (inter-scalar) regional interstitial space that connects local, regional, and even global networks of economic interchange. Regions such as Stockholm in Sweden, Rotterdam in the Netherlands, Guangzhou in China, or those cases in which large portions of water lie between a wider regional urban system of interchange have also environmental salience (Rutherford, 2008). This is the case of maritime space of the Irish sea – between Ireland and Great Britain – which is a daily commuting space that connects several cities and economic hubs such as Dublin, Liverpool, Belfast, Bangor, Douglas (Isle of Man), Glasgow, and others. The Irish Sea is of significant economic importance to regional trade, shipping, and transport, as well as fishing (Phillipson & Symes, 2018) and power generation in the form of wind power and nuclear power plants. This regional interstitial space is inherently heterotopic and has been used as source of gas and oil exploitation, construction of strategic industrial installations, and for the disposal of toxic items from munitions, radiation waste, and domestic and commercial sewage (*The Guardian*, 2020; *The Irish Times*, 1997, 1998, 2019). This space is currently a matter of controversies, considering its condition as natural border but also potentially institutional, framed by the negotiations around the implications of the Brexit and the associated political and economic cost–benefit equations (Anderson, 2018). As such, 'the "Irish Sea border" alternative – at ports and airports in Britain and Ireland – would also have costs (inevitable with Brexit), but in comparison with a very costly and ineffective land border it would be the 'lesser of two evils' or the more efficient, cheaper, damage-limitation option' (ibid, p. 256). This space also connects to remote places and other global interstitial hubs elsewhere, beyond the regional hinterland.

5.2.4 The scale of remoteness

Interstitial spaces of *remoteness* are the largest non-urban geographies between cities and regions. Most of the interstitial spaces defined at this scale are connected to national (and international) development policies and environmental designations such as national parks, protected ecosystems, large geological and topographical features such as mountains, lakes, estuaries. These are also geographical spaces of open countryside or maritime space between countries. As boundary spaces between countries, some interstitial spaces of remoteness are still under dispute and with no clear definition regarding the exploitation of resources or territorial jurisprudence. Some other interstices of remoteness imply wider international agreements in terms of circulations of goods and people. As such, they can be functionally independent and host different interstitial hubs that connect to global trade. This is the case of national parks, protected ecosystems, large magnitudes of countryside and natural landscape, large mining and exploitation sites, economic enclaves, or geographical spaces defined by mountains, lakes, sea channels, and others. Certainly, this scale does not usually illustrate direct morphological impacts on cities, although it has been demonstrated that in some cases cross-national sprawling conurbations take place in these interstitial spaces of remoteness or are influenced by the economic activity embraced by them. It has been observed that 'the emergent process of extended urbanization is producing a variegated urban fabric that, rather than being simply concentrated within nodal points or confined within bounded regions, is now woven unevenly and yet ever more densely across vast stretches of the entire world' (Brenner, 2013, p. 90), which confirms the heterogeneous nature of the scale of remoteness.

There is no doubt that this scale is gaining attention, considering that some of the most radical environmental alterations and economic changes are taking place in the countryside between urban regions. As a global space, the interstices of remoteness describe the nexus between the formation of great global city-regions composed of an extended urbanised area (often comprising more than one metropolis) and a widely ranging hinterland that spans non-urban geographies. From an economic perspective, Mainet and Racaud (2015) highlight the interstices formed by the trading space between small and medium-sized towns in East Africa mountain areas – in Uganda and Tanzania – in which secondary towns act as nodes of wide transnational trade networks. Somehow, this in-between space of global trade has become interstitial, while the small towns and villages that articulate the trading activity – settlements that cannot be labelled as proper cities – become 'interstitial cities' (Saying et al., 2020).

Irrespective of physical distance, while connected to global economic processes this wider geographical space directly (or indirectly) influences the evolution of cities, enclaves, towns and villages around them. Choplin and Pliez (2015) reflect upon the role of the 'invisible transnational connections' that take place in the interstitial spaces of the geography of globalisation. Here, three types of interstitial hubs are identified: routes, secondary cities, and marketplaces. These infrastructural elements connect major and secondary cities, but also other spatial forms that have become quite characteristic of the globalisation and transnational interconnections. The infrastructural character of this context of remoteness has been also highlighted by Rem Koolhaas, who refers to 'the countryside' as the space that lies beyond cities, and in which the magnitude and novelty of the infrastructures, environmental processes, technological innovations, and the size of architectural artifacts have no precedents in human history. The architect indicates that the interstitial space of remoteness is somehow hidden to (or uncontrollable by) the sight of spatial planning conventions and, as such, 'enables a brave new architecture to be realized in record time, without bureaucratic interference' (Koolhaas, 2021, p. 272); infrastructures that are randomly arranged – without connections or shared aim – and with colossal loading bays that make the typical large parking lots that we see in cities something of 'timid size' (p. 272).

The colossal landscapes of the interstices of remoteness are not only infrastructural but also cultural, environmental, and economic. Their significance in terms of impacts is global and describes a temporality that ranges between the few years of obsolescence of an infrastructure up to centuries determined by the geological cycles of climate change or the cultural footprints of indigenous people that have been wandering in these landscapes for centuries (Bocco, 2016). This is the case of the Australian desert, for instance – an interstitial space compressed between the coastal ring of colonial occupation – or the ancient Silk Road space between China and Europe. These cultural and economic dimensions of the interstitial spaces of remoteness have also been drawn from economic geography insights on contemporary planetary urbanisation, which identify systems of cities and hubs that re-conceptualise cross-scale interactions (Ernstson et al., 2010). Aside from the old notion of 'remoteness' – the one that describes isolated and distant places on Earth – it now describes the global space of interdependencies between technical and social networks that connect cities and regions and sustain constant flows of energy, material interchange, and information (Ernstson, 2021). In this scale of remoteness, the interstitial hubs and the space of networks become critical infrastructures (Graham, 2010) that spatialise the interactions between cities and regions through the interstitial spaces of remoteness, finally making remote places less isolated and thus, 'less remote'. The colossal landscapes of networks influence the urbanisation of cities even when the hubs are far afield from where the city is located. This has been profoundly developed by Nicholas Phelps in his book 'Interplaces' (2017), arguing that cities 'coexist with other geographical formations – agglomerations, enclaves, networks, and arenas – as a product of the tension between place and space apparent in the production of space within contemporary capitalism' (p. 40).

It is important to emphasise that the notion of a purely 'rural' realm occupying the interstitial spaces between cities would be archaic and misleading as these geographical magnitudes of global trade describe a varied spectrum of land-uses and geographies beyond mere rurality. Similarly, arguing that these spaces are 'natural' landscapes would be glaringly inadequate as there is enough empirical evidence to assume that – in an era of global capitalist urbanisation – the 'natural' environment is not anymore delinked from the city and its social scope and 'has started to be involved in capital accumulation as an internal factor' (Keil, 2020, p. 2358). Since long ago, the 'wicked problems' linked to ecological modernisation are not only problems of 'cities', and 'cities' are not the unique containers where social processes unfold (Storper & Scott, 2016). The ecological issues linked to societal progress have been extended beyond the realm of cities and, as such, are as social and political as those usually found in cities. The 'natural environment' is no longer external to social processes but indeed influenced and determined by them irrespective of 'where' they occur; 'nature' is currently entangled with a social dimension that cannot be neglected (Keil, 2020). Insights from urban political ecology touch on these issues – occurred in the realm of (planetary) interstitial spaces of remoteness – when suggesting that globally up-scaled infrastructures linked to environmental processes reveal larger spaces of material, ecological, political, and economic production that influence the transformation of whole regional systems (Arboleda, 2020). This scale of remoteness is the space where globalisation takes the form of a rather heterogeneous landscape – contrary to the supposedly homogeneous derivation of global capitalist political economy – and where patterns and processes of production and employment require a more detailed analysis of regionally differentiated places; an analysis that can eventually uncover alternative views of more dynamic and ever-changing places and spaces (Massey, 2005).

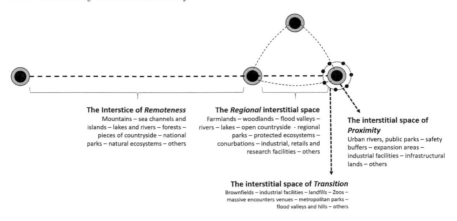

The Interstice of *Remoteness*
Mountains – sea channels and islands – lakes and rivers – forests – pieces of countryside – national parks – natural ecosystems – others

The *Regional* interstitial space
Farmlands – woodlands – flood valleys – rivers – lakes – open countryside - regional parks – protected ecosystems – conurbations – industrial, retails and research facilities – others

The interstitial space of *Proximity*
Urban rivers, public parks – safety buffers – expansion areas – industrial facilities – infrastructural lands – others

The interstitial space of *Transition*
Brownfields – industrial facilities – landfills – Zoos – massive encounters venues – metropolitan parks – flood valleys and hills – others

Figure 5.3 The four scales of interstitiality.

Source: Author.

These four scales of interstitiality are not absolute in the sense that the interstitial spaces and hubs placed in one scale can have transcalar implications and connections to other scales. It has been discussed that an 'interstitial city', for instance, can have salience at both the regional and the scale of remoteness. The same applies to a mining site or a harbour, which can be located in the outer space of a sprawling region while connected to the global hinterland of international trade. These four scales, however, help in locating the geographical position of the interstitial spaces and their empirical tractability (Figure 5.3).

5.3 Conclusions

Interstitial spaces manifest at the scales of proximity, transition, regional, or remoteness. However, the scales at which they manifest are neither absolute nor rigid, as in some cases interstitial spaces have transcalar salience in both spatial and functional terms. The smaller and often more constrained (liminal) interstices of proximity are more likely to represent the piecemeal opportunities distributed within urban regime-style politics and also the tensions found in the intersections of dissimilar communities, imaginaries, and associated politics of the space. The interstices of transition emerge as spaces connected to their surroundings and define their character regarding changes in their boundary areas and potential uses. The regional interstitial spaces connect to wider networks of global interchange while directly influence the urban-rural transect of city-regions. Finally, the scale of remoteness describes the geographical scope of colossal landscapes and hubs that speak about the tensions between locality and globalisation. At this scale, interstitial spaces become autonomous entities that function under their own dynamics. This is the scale in which the city is not producing interstitiality, but rather the interstitiality is influencing the vicissitudes of surrounding cities and far afield places.

It is possible to suggest that the scales of interstitiality have ramifications for extant theories of urban politics. Cities and urbanised regions have been the focus of modernisation despite the fact that a considerable amount of wealth – and somehow ignored transformative societal processes – are nevertheless produced between (and beyond) cities and regions. On this basis, the planning orthodoxies already applied in the studies of cities would need to be revisited in the studies of the interstitiality to incorporate the anthropological and ecological insights that shape the transcalar nature of interstitial spaces. Despite being considered a somehow 'ignored' realm, it has been clarified that the different scales of interstitiality suppose the interaction of different actors, communities, and institutional representations, all elements that finally question the underlying assumption of being a realm truly ignored (or ignored by 'who'). This total (or partial) lack of awareness poses further challenges around the ontologies of comprehensive or fragmented governance of the interstitiality and the studies of their condition as place and space.

References

Arboleda, M. (2020). *Planetary mine. Territories of extraction under late capitalism.* Verso.

Arvanitis, E., Yelland, N. J., & Kiprianos, P. (2019). Liminal spaces of temporary dwellings: Transitioning to new lives in times of crisis. *Journal of Research in Childhood Education, 33*(1), 134–144.

Belfast Harbour Port Authority (2018). *Annual Report & Accounts 2018.* https://www.belfast-harbour.co.uk/corporate/about-us-1 (accessed April 2021).

Bocco, G. (2016). Remoteness and remote places. A geographic perspective. *Geoforum, 77*, 178–181.

Brenner, N. (2013). Theses on urbanization. *Public Culture, 25*(1), 85–114.

Brighenti, A. M. (2013). *Urban interstices: The aesthetics and the politics of the in-between.* Ashgate.

Choplin, A., & Pliez, O. (2015). The inconspicuous spaces of globalization. *Articulo-Journal of Urban Research, 12.* https://doi.org/10.4000/articulo.2905 (retrieved in May 2021).

Douglas, I. (2008). Environmental change in peri-urban areas and human and ecosystem health. *Geography Compass, 2*(4), 1095–1137.

Ernstson, H. (2021). Ecosystems and urbanization: A colossal meeting of giant complexities. *Ambio, 50*, 1639–1643.

Ernstson, H., Van der Leeuw, S. E., Redman, C. L., Meffert, D. J., Davis, G., Alfsen, C., & Elmqvist, T. (2010). Urban transitions: On urban resilience and human-dominated ecosystems. *Ambio, 39*(8), 531–545.

Foo, K., Martin, D., Wool, C., & Polsky, C. (2013). The production of urban vacant land: Relational placemaking in Boston, MA neighborhoods. *Cities, 35*, 156–163.

Gilley, A. (2006). Fractalled: The interstitial spaces and Frank Gehry. Baltimore: Institute of Architecture and Planning Morgan State University. *Bridge Proceedings, 26,* 315–320.

Graham, S. (2010). *Disrupted cities: When infrastructure fails.* Routledge.

Harrison, J., & Heley, J. (2015). Governing beyond the metropolis: Placing the rural in city-region development. *Urban Studies, 52*(6), 1113–1133.

Herrault, H., & Murtagh, B. (2019). Shared space in post-conflict Belfast. *Space and Polity, 23*(3), 251–264.

Iossifova, D. (2013). Searching for common ground: Urban borderlands in a world of borders and boundaries. *Cities, 34*(October), 1–5.

Kabisch, N., & Haase, D. (2013). Green spaces of European cities revisited for 1990–2006. *Landscape and Urban Planning, 110,* 113–122.

Keil, R. (2020). An urban political ecology for a world of cities. *Urban Studies, 57*(11), 2357–2370.

Koolhaas, R. (2021). *Countryside. A report. The countryside in your pocket!* Taschen.

Mainet, H., & Racaud, S. (2015). Secondary towns in globalization: Lessons from East Africa. *Articulo-Journal of Urban Research, 12.* https://doi.org/10.4000/articulo.2880 (retrieved in May 2021).

Maruani, T., & Amit-Cohen, I. (2007). Open space planning models: A review of approaches and methods. *Landscape and Urban Planning, 81*(1), 1–13.

Massey, D. (2005). *For space.* Sage.

Meijers, E. J., & Burger, M. J. (2017). Stretching the concept of 'borrowed size'. *Urban Studies, 54*(1), 269–291.

Morrison, N. (2010). A green belt under pressure: The case of Cambridge, England. *Planning Practice & Research, 25*(2), 157–181.

Nechyba, T., & Walsh, R. (2004). Urban sprawl. *The Journal of Economic Perspectives, 18*(4), 177–200.

Phelps, N. A. (2017). *Interplaces: An economic geography of the inter-urban and international economies.* Oxford University Press.

Platt, R. H. (2004). *Land use and society (revised edition).* Island Press.

Rickards, L., Gleeson, B., Boyle, M., & O'Callaghan, C. (2016). Urban studies after the age of the city. *Urban Studies, 53*(8), 1523–1541.

Roberts, L. (2018). *Spatial anthropology. Excursions in liminal spaces.* Rowman & Littlefield International Ltd.

Rutherford, J. (2008). Unbundling Stockholm: The networks, planning and social welfare nexus beyond the unitary city. *Geoforum, 39*(6), 1871–1883.

Sayın, Ö., Hoyler, M., & Harrison, J. (2020). Doing comparative urbanism differently: Conjunctural cities and the stress-testing of urban theory. *Urban Studies,* 1–18.

Scott, A. J. (2012). *A world in emergence: Cities and regions in the 21st century.* Edward Elgar Publishing.

Steele, W., & Keys, C. (2015). Interstitial space and everyday housing practices. *Housing, Theory and Society, 32*(1), 112–125.

Storper, M., & Scott, A. J. (2016). Current debates in urban theory: A critical assessment. *Urban studies, 53*(6), 1114–1136.

The Guardian. (2020). Records lost of chemicals and arms dumped at sea – archive, 1995. https://www.theguardian.com/uk-news/2020/apr/28/records-lost-of-chemicals-and-arms-dumped-at-sea-archive-1995 (accessed in May 2021).

The Irish Times. (1997). Radioactive waste was dumped in Irish Sea. https://www.irishtimes.com/news/radioactive-waste-was-dumped-in-irish-sea-1.86450 (accessed in May 2021).

The Irish Times. (1998). Sellafield's nuclear pollution of the Irish Sea took decades to achieve. https://www.irishtimes.com/culture/sellafield-s-nuclear-pollution-of-the-irish-sea-took-decades-to-achieve-1.157519 (accessed in April, 2021).

The Irish Times. (2019). Turbulent water: A cultural history of the Irish Sea. https://www.irishtimes.com/culture/books/turbulent-water-a-cultural-history-of-the-irish-sea-1.3864668 (accessed in May 2021).

Titanic Quarter Ltd. (2021). https://titanicquarter.com/live/living/ (accessed April 2021).

Vidal, R. (2002). *Fragmentation de la Ville et nouveaux modes de composition urbaine*. Editions L'Harmattan.

Zhang, H., & Grydehøj, A. (2020). Locating the interstitial island: Integration of Zhoushan Archipelago into the Yangtze River Delta urban agglomeration. *Urban Studies*, 58(10), 2157–2173.

6 The relational character of interstitial spaces

> Instead of thinking of places as areas with boundaries, they can be imagined as articulated moments in networks of social relations.
>
> (Massey, 1991, p. 29)

6.1 Introduction

While connecting different spaces – serving as scenarios of different social and economic interactions and the functional articulation of interstitial hubs – interstitial spaces suggest different levels of relationality. The relationality of interstitial spaces is linked to the infrastructures and networks that connect different hubs, places, and cities but also their functions and magnitudes of the interstices' surroundings. It has been also discussed that the spatial properties of interstitial spaces – whether they are liminal, spaces of transition, or wider geographies between cities and regions – also influence their relational character as they connect (or separate) communities or global systems of economic interchange (Arvanitis et al., 2019).

In this chapter, the relationality of interstitial spaces is clarified and discussed in conceptual and empirical terms based on three aspects: (a) *infrastructures*, (b) *functions*, and (c) *spatiality*, and how they operate as interlinked while defining the relationality of interstices. The relationality of interstitial spaces highlights their condition as in-between entities and the infrastructures that they are part of. As infrastructure-rich spaces, they fall into ambivalent categories of 'placelessness' (Relph, 1976, p. 90) or independent spaces of interchange and development of environmental processes of re-production of the space; non-places that can be quantified 'by totalling all the air, rail and motorway routes, the mobile cabins called "means of transport" (…) and finally the complex skein of cables and wireless networks that mobilize extraterrestrial space for the purpose of a communication so peculiar that it often puts the individual in contact only with another image of himself' (Augé, 2008, p. 64).

DOI: 10.4324/9780429320019-6

6.2 The infrastructures and networks of interstices

Vidal's (1999, 2002) idea of 'interfragmentary space' emerged from a desire to explore the cross-boundary relations between fragments. Vidal's approach is explicit on the potential and relationality of interfragmentary spaces as a result of changes to surrounding fragments and infrastructure of the built environment. This infrastructure supports the creation of networks in which 'the physical distance is not *a sine qua non* condition for networks creation; a communication network could be useful to link fragments even when these could be beside one another, but the interchange is disrupted by an intermediate element. This is the case, for instance, of two neighbourhoods separated by a railway line but connected by telephone lines, a footbridge, a tunnel or a bridge' (Vidal, 2002, p. 157). These infrastructures speak about the multifaceted character of the relationality of interstitial spaces, which when intensified 'displays an increasingly "nested" character, with interacting resource flows, technological interconnections, operational and financial interdependencies and manifold governance interfaces at multiple scales' (Monstadt & Coutard, 2019, p. 2192).

On a metropolitan scale, the planning relevance of interstitial spaces is determined by their transport infrastructure (or proximity to); a critical factor to evaluate land capacity and how much density an area can support for economic returns (Kombe, 2005; McDonald & Brown, 1984). The land capacity has been subject of debate – not only related to the land size – considering that small interstitial spaces can also be relevant if they have good locations or if regulations allow land-use changes (Monclús, 2003). In some cases, well-served interstitial spaces become important connectors, spaces that will remain undeveloped but intended for connectivity (Graham, 2000). These interstitial spaces become 'network spaces' defined by their capacity to connect other areas. Conurbation zones are clear examples of network spaces, regional interstices crossed by motorways and railways that connect cities with outer developments (Batten, 1995). These interstitial spaces can include a certain degree of development, can be located within the city but can nevertheless continuously extend beyond the city's boundaries. The heightened mobility associated with such infrastructures can appear 'to involve a number of absences – the absence of commitment and attachment and involvement – a lack of significance' (Cresswell, 2006, p. 31). Yet, in contrast to the absences associated with infrastructures of mobility, Augé's discussion on non-places confirms the transitional character of the interstices while 'passing through'; interstices as spatial-temporal instances that can emerge as independent spaces. This independency is reinforced by fragments that are physically separate but connected by lines, channels, mains, pipes, and other elements concerned with promoting movement that take the form of distinctive networks regarding their surroundings (Vidal, 2002).

On a regional scale, these infrastructures carry a negative connotation on interstices as spaces that emerge as physical divisions and restricted areas. Industrial facilities, motorways, railways, and other heavy infrastructure divide places while intensive provision for mobility might signal significant regional economic potential. These infrastructural spaces are nevertheless perceived as negative because they do not afford easy access to pedestrians and to nearby employment sites (Graham & Marvin, 2001). In many instances, the networks can contribute to the indeterminacy of interstitial spaces since infrastructure describes different degrees of functionality and effectiveness coexisting in the same interstitial space. A motorway, for instance, might be seen as an efficient and modern infrastructure coexisting with the now less functional railway as the heritage of the former prime mode of transport of commodities. These two share an interstitial space in which a polluted canal is used to evacuate industrial residues but hardly offers a safe, let alone amenable, artery for connection (Figure 6.1). These heavy infrastructures can transform the interstices into derelict spaces or areas with a certain degree of erosion that impacts the overall environment (Bruinsma, et al., 1993; Relph, 1976). However, the presence of heavy infrastructures can also mean a benefit if interstitial

Figure 6.1 Different networks mingle at the boundary area of Cerrillos and Pedro
Aguirre Cerda communes, Santiago de Chile.

Source: Author, 2014.

spaces are properly recovered, considering that 'the strategic reuse of urban vacant land and abandoned structures can represent a key opportunity for encouraging greater density and reducing the push to develop suburban greenfields' (Pagano & Bowman, 2000, p. 1). In any case, this amalgamation of infrastructures coexisting within the same interstitial space speaks about the politics of integration, cross-coordination, and approaches that critically look beyond individual domains. The overlapped infrastructure defines a relationality of cross-sector interactions, interfaces, and hybridities that are essential for a broader understanding of the nature of interstitial spaces and the way of how they contribute to the contemporary urban condition.

In the context of interstitial spaces of remoteness, 'the spatial expansion of large networks has supported the advent of "generalised" or "planetary" urbanisation' (Monstadt & Coutard, 2019, p. 2200). Here, the supposedly pristine and untouched natural world is also a source of natural resources that become commodified and integrated into the city through technological networks; again, expanding the planetary urban fabric beyond the boundaries of cities and regions (Brenner & Schmid, 2014). The relationality of the interstitial spaces of remoteness codify (and commodify) nature to the extent to which nature is being 're-invented in its urban form' (Kaika & Swyngedouw, 2000, p. 121). So, wider networks of pipes, electric lines, conduits through which water, energy, raw material, financial capital, technology, and information become part of dense metabolic flows that are politically mediated. These infrastructures go across thousands of kilometres over the rural fields as visible expressions of material interchange but also sometimes simply hidden from view (Kaika, 2005). Their implied relationality defines cities as more connected to a planetary system of production than to their direct surroundings, and thus, the urban condition from the scale of remoteness should no longer be understood in terms of binary urban-rural relations but in terms of more or less extended forms of urbanisation (Brenner & Schmid, 2014). In this context, infrastructures of relationality connect (and support) ecological, financial, operational, and institutional interactions and overlap different independent infrastructures. This suggests a hybridisation between the different domains that cannot manifest themselves as separated and autonomous. The infrastructures of the interstitiality 'rely on each other and coevolve in their interrelationships with urban development, and are interwoven within the fabric of urban space' (Monstadt & Coutard, 2019, p. 2194).

This hybridisation of infrastructures has not been indifferent from the sight of planners and designers. Perhaps, one of the ultimate utopias about hybridisation of networks and interstitial spaces can be seen in Le Corbusier's plans for Algiers. In 1933, the architect presented his 'Plan Obus' in which a long arching roadway shows the double function of being a road and a housing block simultaneously – all amalgamated together in the same morphological element. This is literally a 'viaduct city' proposed to connect central Algiers to its suburbs and the curvilinear complex of housing in the heights that accesses the waterfront business district via an elevated highway bypassing the Casbah. Housing and commercial functions were located above and below the great route, which thus became a linear city: a space of movement while it is also a place to live (Morshed, 2002). Another example – previous to Le Corbusier's Plan Obus – is Edgar Chambless's 'Roadtown' (1910), in which 'the urban fabric follows the railway right-of-way and inhabits a mixed-use envelope around it' (Savvides, 2004, p. 51). Aside from utopias, some of these compositions are possible to be seen today in many cities, although their visibility is often only a matter of touristic attraction due to their peculiar morphological configuration. This is the case of the Ponte Vecchio (Vecchio Bridge) in Florence, in which a whole residential and commercial block is literally 'the bridge' over the Arno River (Figure 6.2).

The relationality of interstitial spaces defined by their infrastructures also resonates with their environmental character. Narratives on green and blue infrastructure in some way situate the interstitial spaces themselves as (green/blue) infrastructural spaces, but also the infrastructure that built their relationality. As the networks have become the metabolic systems of the interstices (and cities), it has been demonstrated that

Figure 6.2 The Ponte Vecchio over the Arno River, Florence.

Source: Author, 2016.

when they fail, they trigger dramatic social, economic, and environmental crises. In that sense, the relationality of interstitial spaces become critical on issues of urban resilience (Meerow & Newell, 2019), and the way of how urban system address their recovery after extreme weather events or other disruptions (Matthews et al., 2015). The terms green and blue infrastructure carry a sense of how interstitial spaces are forces for the integration of urban territories (Barbati et al., 2013; Kim et al., 2015). The infrastructural properties of the interstices as open and green space are also implicit in Sieverts' emphasis on the conflicts over major public works in the *Zwischenstadt* (2003). This overlaps of governance and meanings suggest a transcalar management and relations between cities and interstitial spaces; an inter-scalar reality of the network space that also has non-physical salience when it becomes culturally and socially meaningful: 'from the intimate space of the bathroom to the hidden space of the sewer, from the civic display of the fountain to the distant dams and the reservoir, water provides a link between the corporeal experience of space and the abstract dynamics of capitalist urbanization' (Gandy, 2004, p. 178).

6.3 The functional relationality of interstitial spaces

Interstitial spaces can also define their relational character in regard to their functions. A military facility located outside a city, for instance, can be a large open space intended for military practices and research. This is a spatially open space but functionally closed – with restricted access to the public – and with limited exposure of its land management and internal modes of governance (Harmon et al., 2014). As a restricted space, its relational character becomes limited, above all regarding its immediate surroundings. However, its modes of interaction are determined by its location within the regional and national space. So, a military space appears functionally isolated from its local hinterland but connected to wider regional and national space. A similar situation applies to similar interstitial hubs with restricted access such as research centres, business corporative lands, large industrial hubs, mining sites, and agricultural research centres (Figure 6.3). These interstitial spaces become functional 'islands', isolated spaces that illustrate a dual relationality: while showing a very limited relationality with their scales of proximity, they show a fluid relationality at regional and national scales.

Zhang and Grydehøj (2021) analyses the case of the Zhoushan Archipelago (Zhejiang Province, China) as an interface zone both between cities within the Yangtze River Delta urban agglomeration and between the Yangtze River Delta urban agglomeration and other megaregions. The authors contest the notion of 'island' as mere pieces of land surrounded by water – linked to remoteness, isolation, and otherworldliness – as contemporary insights from urban island research have 'emerged in interaction with a "relational turn" in island studies' (p. 2161). The study

Figure 6.3 An extraction site in the commune of Cerrillos. Santiago de Chile.

Source: Author, 2014.

demonstrates that the Zhoushan Archipelago is not 'peripheral' in the sense that it fully contributes to significant economic, industrial, and transport functions that Zhoushan provides for its mainland hinterlands at both regional and global scales. This role is precisely based on the relational character of the island as 'Zhoushan – which has for so many centuries been simultaneously within and outside the urban worlds of mainland China, Korea, Japan, and now the Pacific as a whole – continues to be regarded as a distinct place precisely because of (rather than despite) its connections with other places' (p. 12). Similarly, Chandler and Pugh (2020) describe the role of islands as part of complex networks of relations, assemblages, and flows; islands are relational spaces 'that unsettle borders of land/sea, island/mainland, and problematise static tropes of island insularity, isolation, dependency and peripherality' (p. 65). In a different vein, interstitial spaces of restricted relationality with their immediate surroundings but functionally connected to their regional and global hinterland are the emerging privatised ecological reservoirs. Although there is enough evidence to suggest that many privatised conservation areas do not coincide with concrete ecologically significant areas (Martelli, 2021), their relationality is clear: while they remain closed to the public and other functional requirements from government agencies, they remain largely connected to the environmental processes they are part of (Gallo et al., 2009).

Other functions related to farming, entertainment parks, ecological reservations or temporary uses do not demand functional isolation and thus, allow a certain level of controlled relationality with their surroundings (Németh & Langhorst, 2014). These are the cases of some derelict suburban lands informally transformed into sports pitches, recreational or public spaces, or ecological reservoirs used for educational, touristic, or social purposes (Silva, 2020). So, the relational character of interstitial spaces is determined by the combination of spatial and functional aspects.

6.4 The spatial relationality of interstitial spaces

Spatial aspects of interstitial spaces also influence their level of relationality, ranging from totally opened to totally closed spaces. A farming space, a valley, a lake, or a hill embedded within the sprawling fabric of city-regions would only be accessible if their boundaries and surroundings present a certain level of permeability that allows connections. Natural fences such as surrounding forests, water trenches, or hills would appear as barriers that diminish their relationality. Similarly, artificial boundaries such as fences, a ring road, or walls would operate in a similar way diminishing the relationality of an interstitial space. Alternatively, the relationality would be increased if boundaries are penetrable or if an infrastructure of penetration is provided (a bridge, a road, or pathways that connect the interior and exterior sides of the interstice). Boundaries are socially constructed and serve to minimise ambiguity around the rights and ownership of space (van Houtum & Van Naerssen, 2002). In that sense, they can be closed and rigid, or open and permeable, facilitating the socio-spatial interchange between interstitial spaces and hubs. Thus, by changing the boundaries – or by providing an infrastructure of penetration – the relationality of an interstitial space can be modified.

The notion of 'boundary' has been widely used in the disciplines of the built environment, human geography, sociology, and philosophy (Uchiyama & Mori, 2017; Vis, 2018). Detailed work on 'boundaries' concerning the urban form is developed by Benjamin Vis (2018), who links the notion of boundaries with the social processes that unfold in cities. Focussing on the built environment, the author argues that 'built boundaries compose the built environment as outcomes of transformative differentiating interactions', so the 'urban form is then seen as a configuration of boundaries' (pp. 99–100). Although these theoretical insights are 'city-centred' – and distilled from observations on the built-up space (accepted here as the 'urban form') – the conceptual constructions would apply for the case of interstitial spaces. The author indicates that boundaries suppose 'differentiation' as 'we come to know any entity by its distinction to its outside' (ibid, p. 102). Somehow, boundaries are social or regulatory constructions that can be unmade despite their material composition as elements that allow the recognition of interstitial spaces as such. Boundaries – as elements of the urban form – and the interstitial space are imbricated entities in which one cannot exist without the other. Vis (2018) argues that 'how entities become physically distinguished by boundaries is the specific constitutive human and social datum of the inhabited built environment' (p. 102). A closely related literature on 'borders' indicates that boundaries can take the form of borders, and as such, they illustrate a certain degree of autonomy – differentiated as spaces in-between – and distinguished from self-contained urban enclaves or other forms of urbanisation with politically defined forms of governance (Iossifova, 2013).

Star's work on 'boundary objects' (1989) also moves the reflection on how boundaries can be framed for the interstitial spaces. Boundary objects are elements that act as platforms for standardised knowledge and interchange of information that can then be used by different communities, groups, actors, or individuals that share common goals. The concept alludes to the creation of common grounds between different fields and politically mediated representations that enable interdisciplinarity but also cooperation, scientific consensus, and 'a mutual *modus operandi*' (Star & Griesemer, 1989, p. 388). This would apply to the mediating spaces, devices, infrastructures, and surroundings that confine the interstitial spaces in the sense that 'boundary objects' are 'objects which both inhabit several intersecting social worlds (...) and satisfy the informational requirements of each of them' (ibid, p. 393). The authors also indicate that 'boundary objects' are 'plastic enough to adapt to local needs and the constraints of the several parties employing them' (ibid, p. 393), which speaks about the adaptability of boundaries to improve the relationality of interstices. In this case, the 'boundary' of interstitial spaces would be more an enabling device, a bridge space that allows communication between spaces and actors which converge in the interstitial space. It can also be inferred that interstitial spaces can be boundary objects themselves, as spaces that enable the interaction of different actors and institutions once they become a matter of common interest. London's green belt can be an interstitial space that operates as a boundary object between the city and the rural and environmental representations that lie beyond the city's limit (Dockerill & Sturzaker, 2020). Moreover – and as discussed – interstitial spaces can be circumscribed by other boundary objects that would confirm their interstitial condition and define their relationality regarding what lies beyond. Different from Star's notion of boundary objects as bridging elements, boundary objects can or cannot be 'relational' and thus, operate as 'barriers' rather than 'bridges'. This is the case of a suburban landfill surrounded by motorways that impede the connection of the landfill with the urban fabric. The motorways are hard boundaries that leave the site disconnected and isolated, the boundary objects (the motorways) that diminish the relationality of the interstitial space (the landfill). These are the cases in which 'boundary crossing' can be introduced as modes of entering into a territory we are unfamiliar with (Akkerman & Bakker, 2011) and 'face the challenge of negotiating and combining ingredients from different contexts to achieve hybrid situations'

(Engeström et al., 1995, p. 319). The understanding of interstices' surroundings as 'boundary objects' subjects of 'boundary crossing' is utilitarian to indicate how spatial artifacts can fulfill a specific function in bridging intersecting practices (Akkerman & Bakker, 2011). For Vidal, boundaries (and surrounding) help to define the character and identity of an interstitial space, and thus, the interstitial space can be the realm in which different identities converge; a space of individual and collective significance that can change its nature regarding changes of its boundaries (Vidal, 2002) and so, its relationality.

Considering the varied configurations of boundaries, interstitial spaces with an apparently very low degree of relationality can connect with other interstitial spaces, hubs, cities, and regions in one way or another. While physical boundaries can be apparently impenetrable from a functional perspective, their institutional relations can be fluent and linked to wider societal processes. On this basis, the spatial condition of interstices regarding their relationality is the one of 'more or less' (relational), and boundaries speak more about the balance between the internal and external social, economic and political dynamics that affects the interstitial spaces. As 'spaces' the interstices are 'articulated moments in networks of social relations' (Massey, 1991, p. 29), where politics of the urban define the relationality of interstitial spaces as product of interrelations at different scales and through different modes: *infrastructures*, *functions*, and *spatialities*; spaces that are 'constituted through interactions, from the immensity of the global to the intimately tiny' (Massey, 2005, p. 9).

6.5 Conclusions

The relationality of interstitial spaces is determined by the infrastructures and networks, functions, and spatial elements that compose their interstitial condition. Infrastructures can penetrate the interstices, move across, create network spaces composed of different infrastructures, and extend beyond the boundaries of the interstices to connect other interstitial spaces, hubs, cities, and regions. The infrastructure supports the mobility of goods and people but also the development of social and political processes of transformation. The functions of the interstitial spaces can also determine their level of relationality as spaces with restricted access that appear closed to their immediate surroundings while simultaneously connected to the wider regional and global networks of economic exchange. Conversely, some interstitial spaces can describe functions that connect with their immediate surroundings while appear disconnected (or less relational) from the dynamics of the regional space. Finally, the relationality of interstitial spaces is determined by their spatial character, in which boundaries play a critical role as bridge elements that connect interstices with their surroundings at different levels and scales. Strategies of 'boundary crossing' can be defined, however, to explore the ways of how interstitial spaces can be further integrated, a boundary context where intersections of political, economic, and cultural practices open up third spaces – the boundaries of interstices – that allow negotiation and functional hybridity.

The relationality of interstitial spaces can be defined by a combination of these infrastructural, functional, and spatial aspects that calibrate the extent to which they become more or less relational – or more autonomous – within the interstitial geography or within cities. As such, the relationality of interstitial spaces can vary, be multifaceted, and possibly be controlled by adjusting some of the aspects that affect their relationality; eventually a device of 'boundary crossing' – in the form of another infrastructure – can be introduced to modify the relational character of an interstitial space. Further research would be needed, however, around the boundary objects and boundary crossing to explore how they influence the relationality of interstitial spaces.

References

Akkerman, S. F., & Bakker, A. (2011). Boundary crossing and boundary objects. *Review of Educational Research*, *81*(2), 132–169.

Arvanitis, E., Yelland, N. J., & Kiprianos, P. (2019). Liminal spaces of temporary dwellings: Transitioning to new lives in times of crisis. *Journal of Research in Childhood Education*, *33*(1), 134–144.

Augé, M. (2008). *Non-places. An introduction to supermodernity*. Verso.

Barbati, A., Corona, P., Salvati, L., & Gasparella, L. (2013). Natural forest expansion into suburban countryside: Gained ground for a green infrastructure? *Urban Forestry & Urban Greening*, *12*(1), 36–43.

Batten, D. (1995). Network cities: Creative urban agglomerations for the 21st century. *Urban Studies*, *32*(2), 313–327.

Brenner, N., & Schmid, C. (2014). The 'urban age' in question. *International Journal of Urban and Regional Research*, *38*(3), 731–755.

Bruinsma, F., Pepping, G., & Rietveld, P. (1993). Infrastructure and urban development: The case of The Amsterdam orbital motorway. *Serie Research Memoranda*. Faculteit der Economische Wetenschappen en Econometrie. Vrije Universiteit Amsterdam.

Chambless, E. (1910). *Roadtown*. Roadtown Press.

Chandler, D., & Pugh, J. (2020). Islands of relationality and resilience: The shifting stakes of the Anthropocene. *Area*, *52*(1), 65–72.

Cresswell, T. (2006). *On the move: Mobility in the modern western world*. Routledge.

Dockerill, B., & Sturzaker, J. (2020). Green belts and urban containment: The Merseyside experience. *Planning Perspectives*, *35*(4), 583–608.

Engeström, Y., Engeström, R., & Kärkkäinen, M. (1995). Polycontextuality and boundary crossing in expert cognition: Learning and problem solving in complex work activities. *Learning and Instruction*, *5*, 319–336.

Gallo, J. A., Pasquini, L., Reyers, B., & Cowling, R. M. (2009). The role of private conservation areas in biodiversity representation and target achievement within the Little Karoo region, South Africa. *Biological Conservation*, *142*(2), 446–454.

Gandy, M. (2004). Water, modernity and emancipatory urbanism. In L. Lees (Ed.), *The emancipatory city? Paradoxes and possibilities* (pp. 178–191). Sage Publications.

Graham, S. (2000). Constructing premium network spaces: Reflections on infrastructure, networks and contemporary urban development. *International Journal of Urban and Regional Research, 24*(1), 183–200.

Graham, S., & Marvin, S. (2001). *Splintering urbanism: Networked infrastructures, technological mobilities and the urban condition*. Psychology Press.

Harmon, B. A., Goran, W. D., & Harmon, R. S. (2014). Military installations and cities in the twenty-first century: Towards sustainable military installations and adaptable cities. In I. Linkov (Ed.), *Sustainable cities and military installations* (pp. 21–47). Springer.

Iossifova, D. (2013). Searching for common ground: Urban borderlands in a world of borders and boundaries. *Cities, 34*, 1–5.

Kaika, M. (2005). *City of flows: Modernity, nature, and the City*. Routledge.

Kaika, M., & Swyngedouw, E. (2000). Fetishizing the modern city: The phantasmagoria of urban technological networks. *International Journal of Urban and Regional Research, 24*(1), 120–138.

Kim, G., Miller, P. A., & Nowak, D. J. (2015). Assessing urban vacant land ecosystem services: Urban vacant land as green infrastructure in the City of Roanoke, Virginia. *Urban Forestry & Urban Greening, 14*(3), 519–526.

Kombe, W. J. (2005). Land use dynamics in peri-urban areas and their implications on the urban growth and form: The case of Dar es salaam, Tanzania. *Habitat International, 29*(1), 113–135.

Martelli, F. (2021). Buying=saving: Doug Tompkins's Chile campaign. In R. Koolhaas (Ed.), *Countryside? A report. Countryside in your pocket!* (pp. 252–269). Guggenheim/Taschen.

Massey, D. (1991). A global sense of place. *Marxism today*, June (pp.24–29).

Massey, D. (2005). *For space*. Sage.

Matthews, T., Lo, A. Y., & Byrne, J. A. (2015). Reconceptualizing green infrastructure for climate change adaptation: Barriers to adoption and drivers for uptake by spatial planners. *Landscape and Urban Planning, 138*, 155–163.

McDonald, G. T., & Brown, A. L. (1984). The land suitability approach to strategic land-use planning in urban fringe areas. *Landscape Planning, 11*(2), 125–150.

Meerow, S., & Newell, J. P. (2019). Urban resilience for whom, what, when, where, and why? *Urban Geography, 40*(3), 309–329.

Monclús, F. J. (2003). The Barcelona model: And an original formula? From 'reconstruction' to strategic urban projects (1979–2004). *Planning Perspectives, 18*(4), 399–421.

Monstadt, J., & Coutard, O. (2019). Cities in an era of interfacing infrastructures: Politics and spatialities of the urban nexus. *Urban Studies, 56*(11), 2191–2206.

Morshed, A. (2002). The cultural politics of aerial vision: Le corbusier in Brazil (1929). *Journal of Architectural Education, 55*(4), 201–210.

Németh, J., & Langhorst, J. (2014). Rethinking urban transformation: Temporary uses for vacant land. *Cities, 40*, 143–150.

Pagano, M., & Bowman, A. (2000). *Vacant land in cities: An urban Resource*. The Brookings Institution – Survey series (pp. 1–9). Brookings Center on Urban & Metropolitan Policy.

Relph, E. (1976). *Place and placelessness*. Pion.

Savvides, A. (2004). Regenerating urban space: Putting highway airspace to work. *Journal of Urban Design*, *9*(1), 47–71.

Sieverts, T. (2003). *Cities without cities. An interpretation of the Zwischenstadt.* Spon Press.

Silva, C. (2020). The rural lands of urban sprawl: Institutional changes and suburban rurality in Santiago de Chile. *Asian Geographer*, *37*(2), 117–144.

Star, S. L. (1989). The structure of ill-structured solutions: Boundary objects and heterogeneous distributed problem solving. In L. Gasser, & M. N. Huhns (Eds.), *Distributed artificial intelligence* (Vol. *II*, pp. 37–54). Morgan Kaufmann Publishers, Inc.

Star, S. L., & Griesemer, J. R. (1989). Institutional ecology, translations' and boundary objects: Amateurs and professionals in Berkeley's Museum of vertebrate zoology, 1907-39. *Social Studies of Science*, *19*(3), 387–420.

Uchiyama, Y., & Mori, K. (2017). Methods for specifying spatial boundaries of cities in the world: The impacts of delineation methods on city sustainability indices. *Science of the Total Environment*, *592*, 345–356.

van Houtum, H., & Van Naerssen, T. (2002). Bordering, ordering and othering. *Tijdschrift voor economische en sociale geografie*, *93*(2), 125–136.

Vidal, R. (1999). Fragmentos en tensión: Elementos para una teoría de la fragmentación urbana [Fragments in tensión: Elements for a theory of urban fragmentation]. *Revista Geográfica de Valparaíso*, 29(30),149–180.

Vidal, R. (2002). *Fragmentation de la Ville et nouveaux modes de composition urbaine* [Fragmentation of the city and new modes of urban composition]. L'Harmattan.

Vis, B. (2018). *Cities made by boundaries. Mapping social life in urban form.* UCL Press.

Zhang, H., & Grydehøj, A. (2021). Locating the interstitial island: Integration of Zhoushan archipelago into the Yangtze River delta urban agglomeration. *Urban Studies*, 58(10) 2157–2173.

7 Exploring seven interstitial spaces in Santiago de Chile

7.1 Introduction

Latin America is one of the most urbanised regions worldwide, having almost 90% of its population living in cities (UN, 2018). Here, suburban sprawl has become a common pattern of urban development characterised by an extended low-density suburbia, high-levels of land fragmentation, expansion of transport infrastructure, and a wide range of land uses resulting from the extension of cities to the outskirts while spanning industrial facilities, satellite towns, conurbation zones, retail facilities, office parks, farming lands, geographical accidents, and others (Flores et al., 2017; Herzog, 2015). Not only observed in heavily urbanised capital cities, but urban sprawl is also a 'well-established and common phenomena both in large megacities and in the secondary (or even smaller) cities' (Heinrichs & Nuissl, 2015, p. 216), and has included the emergence of informal settlements and slums resulting from massive rural-urban migrations during the last fifty years (Sabatini & Salcedo, 2007). It has been also noticed that the Latin American version of urban sprawl has been subject of the emergence of various large-scale 'gated' residential complexes; clusters of high and middle-income families that share common infrastructure, facilities, security, and are 'separated from the public by a gate, fence or wall' (Heinrichs & Nuissl, 2015, p. 219). The consensus indicates that rapid urbanisation driven by the housing debate has been one of the main drivers of the Latin American version of urban sprawl, partly fuelled by economic growth but also an active participation of the states in delivering massive subsidies to low-income families to be located in new homes at the periphery of cities (Borsdorf & Hidalgo, 2010). This amalgamation has depicted a suburban landscape where 'it appears that by "coincidence" differences in social class in Latin America are currently being play-out in suburbia' (Heinrichs & Nuissl, 2015, p. 220). As such, urban sprawl has had important consequences for socio-spatial segregation (Borsdorf et al., 2007; Heinrichs et al., 2011).

In this context, interstitial spaces have become a common element of the sprawling geography of Latin American cities. These interstices are

DOI: 10.4324/9780429320019-7

quantitatively significant and may be on the increase in contexts of urban shrinkage (Dubeaux & Sabot, 2018). Only the vacant lands of Rio de Janeiro define around 44% of the total municipal area, while it is 21.7% in Quito. El Salvador shows 4.65% of vacant sites, although the authorities also count informal occupations that are included in the list to be demolished, reaching a total of 40% of the whole urban space. In Buenos Aires, vacant lands represent 32% of the metropolitan area (Clichevsky, 2007), while in Santiago de Chile, they signify close to 19%. Similar reports indicate an average closer to 15% in American cities (Cámara Chilena de la Construcción, 2012). These percentages are mainly concentrated in suburban, peri-urban, and fringe-belts areas – all constitutive of urban sprawl contexts. This poses important questions regarding the unsustainable character of urban sprawl – as despite coming under severe criticism – it nevertheless represents a major source of infrastructural and environmental assets (Gavrilidis et al., 2019) and appears as an interface that facilitates exchange between and within vast urban systems (Sieverts, 2003).

The analysis presented in this book is based on urban sprawl in the capital city of Chile, Santiago. This is a metropolitan region composed of 36 communes – each one with its own local authority – and with a population of 7.112.808 inhabitants that represents 40.5% of the country's population (INE, 2017). The metropolitan area describes a monocentric city-model with at least four geographically recognised zones of urban expansion: (a) the *historical city*, (b) the *consolidated city*, (c) the *suburban*, and (d) the *exurban*; the suburban and the exurban comprise the sprawling geography of Santiago clearly delimitated by the Americo Vespucio ring road, which connects most peripheral communes and separates suburban sprawl from the consolidated city (Figure 7.1).

The sprawling geography of Santiago is characterised by built-up lands of different sorts including large-scale housing developments (for low-, middle-, and high-income families), a few peripheral centralities that include high-streets, warehouses, shopping malls, and civic centres that suggest emerging patterns of polycentrism (Truffello & Hidalgo, 2015), and a broad array of interstitial spaces (Gainza & Livert, 2013; Silva, 2019; Silva, 2020; Phelps & Silva, 2017). In the spectrum of interstitial spaces, those selected for analysis are considered strategic by different actors regarding their land capacity, proximity to transport infrastructure, and lack of planning restrictions. These interstitial spaces are mostly located in the south metropolitan area as it is recognised as the most important axis for suburbanisation. This area has been the subject of fragmented sprawling expansion and is a good example of the significance of normative planning and socio-political acceptance of extended suburbanisation (de Mattos et al., 2014).

As context-dependent elements – and as previously discussed – the fortunes of interstitial spaces are linked to the fortunes of the sprawling processes in which they manifest. Therefore, to understand the emergence, role, characteristics, and implications of Santiago's interstitial spaces it is

The 'Orbital' ring-road
(planned, not implemented)

THE HISTORICAL CITY

THE CONSOLIDATED CITY

Américo Vespucio ring-road

THE SUBURBAN CITY

THE EXURBAN

Figure 7.1 Map of Santiago, its four stages of urban growth and boundaries.

Source: Author.

also necessary to analyse the context of Santiago's urban sprawl; determinants that give credit to the proliferation of both built-up and interstitial spaces alike. After this, the selected interstitial spaces are analysed regarding their spatial qualities, meanings, scales, and relationality, to unveil how they define the heterogeneity and the environmental significance of the interstitial geography of Santiago de Chile. Santiago's interstitial spaces stimulate the imaginary of people, politicians, developers, communities, authorities, academics, planning and design professionals, and environmentalists; imaginaries that finally speak about the politics behind the production of the (interstitial) space.

7.2 Urban sprawl in Santiago de Chile

Initial debates around Santiago's sprawl started around the 1980s – after the enactment of The National Policy of Urban Development (NPUD/1979) – which declared that urban land was a non-scarce resource followed by the abolition of the 'urban limit' (Vergara & Boano, 2020).

The latter was a containment policy that divided the urban areas from the rural realm. In several empirical studies around Santiago's urban developments it has been clarified that underlying politics behind the NPUD related to a radical implementation of a neoliberal approach in planning whereby land was commodified and managed under market-driven dynamics (Barton et al., 2012). The NPUD/1979 was revised in 1985, although the essential features of the 1979 version remained, except for emphasising the role of the state in promoting urban development through a subsidiary approach (Gross, 1991). In 2010, a new government commission elaborated a new national policy published in 2014 by the Ministry of Housing and Urbanisation (MINVU). Here, explicit statements to internalise the impacts of the urban expansion are indicated via compensation and promotion of inner densification. In Section 2 (economic development), objective 2.3, point 2.3.5, the policy aims 'to establish a framework of special obligations to new areas of urban expansion, in order to ensure their responsibility on the externalities. Also, to establish a framework of incentives to densification projects to address the impacts over the public space and urban functions in accordance to the law' (MINVU, 2014b, p. 35).

In the NPUD/2014 it is clarified that policies of retrofitting would contribute to sustainable development as an alternative to urban sprawl. This is highlighted by some census data around increments in the housing stock over the past 20 years characterised by a higher construction rate of apartments in comparison to detached houses and a reduction in the rate of expansion in major cities. However, these are minor changes considering that the sprawling pattern of urban development has been a constant trend with permanent impacts at environmental, economic, and social levels (Romero & Órdenes, 2004). There is also a sustained assumption that sprawling growth leads to environmental, residential, and socio-spatial segregation (Romero et al., 2012; Sabatini et al., 2009). This assumption has been extensively examined in empirical studies on housing policies and urban growth (Boccardo, 2011; Borsdorf et al., 2016; Cox & Hurtubia, 2021; de Mattos et al., 2014; Truffello & Hidalgo, 2015) to understand the causes of residential segregation and impacts of Santiago's sprawl in the overall environmental sustainability (Fuentes & Pezoa, 2018; Romero et al., 2012; Romero & Órdenes, 2004). Efforts against 'segregation' have been the focus of the urban debate in Chile in the last years, in which 'the concern about a "social divide" and segregation between groups with different socio-economic status is very visible in the (extant) literature. This comes as no surprise, given that Latin American cities exhibit a high degree of segregation' (Heinrichs & Nuissl, 2015, p. 221). This segregation is recognised by NPUD/2014 as a major challenge 'caused by decades of advances in reducing the housing shortage from a quantitative approach, without paying attention to the location of housing and access to basic public goods' (MINVU, 2014b, p. 9). Although the intention of the policy is clear, the link between locational approaches with the provision of public goods and segregation is still ambiguous as these variables do not directly correlate to each other (Sabatini, 2015), and the policy itself lacks the legal status and mechanisms to control capital (and windfall) gain that strongly influences uncontrolled urban growth (Trivelli, 2015).

Often invoked as 'urban dispersion' (Heinrichs et al., 2009), 'urban fragmentation' (Link, 2008), 'metropolitan expansion' (de Mattos, 1999), 'extended suburbanisation' (de Mattos, 2001), and 'dispersed urban expansion' (Ducci & Gonzalez, 2006), there is now a certain consensus that Santiago's urban development illustrates clear patterns of 'urban sprawl' (Rojas et al., 2013). In a detailed work, Silva and Vergara (2021) identified at least *thirteen determinants* that operate as interlinked at the point of defining Santiago's sprawling geography as a distinctive space that poses significant challenges for urban governance. The determinants are context-dependent, multiple and interlinked, and specifically defined by legal norms, historical relocation of slums, institutional asymmetries between rural and urban agencies, restrictive controls on urban growth, the creation of planned outer developments in the form of gated communities, weaknesses in legal planning norms, specific laws that promote unregulated residential developments outside the urban limits, restrictions to inner densification, improvements in regional connectivity, as well as conceptual ambiguities around the patterns of urban expansion in Santiago. All these determinants 'are placed at different institutional levels – and with repercussions at different spatial scales – configuring a challenging policy context that calls for the development of theories of urban politics beyond normative rationales' (ibid, p. 28).

7.3 The emergence of interstitial spaces of Santiago

Santiago's sprawl describes a range of interstitial spaces recognised by planners, policy-makers, developers, politicians, residents, and the specialised literature as both marginal and also valuable. The types of the interstitial spaces found in Santiago include agricultural and industrial lands, brownfields, landfills, public spaces, geographical restrictions, conurbation zones, former airports, military facilities, small-scale farming areas, research centres, infrastructural spaces, and buffers of security. Some of them are currently well located near transport, energy supply, services, and populated surroundings that make them attractive for both public and private investments.

They are marginal as they are still found outside planning regimes or simply undeveloped. Simultaneously, they are synonyms of spatial diversity and opportunities to change suburban inertias of low-quality urbanisation and socio-residential segregation. However, the meanings of Santiago's interstitial spaces are not the same for different actors and institutional representations. For central authorities and developers, interstitial spaces are 'wasted lands' – in terms of their participation in the land market [interview number 16] – while for policy makers these are spatial disruptions in the urban fabric that need to be converted into built-up areas [interview number 17]. For politicians at central level, interstitial spaces exist as a result of a lack of political will to integrate them into the city's fabric [interview number 20; 21; 53]. For local planners, developers and scholars, interstitial spaces are 'opportunities' to improve urban standards at different levels and reservoirs for services and workplaces [interview number 28; 32]. For residents, Santiago's interstices represent hope but also uncertainty and contradiction [interview number 49; 50; 51]. Somehow, they resemble the rurality within the city while simultaneously being unsafe and derelict (Figure 7.2). The interstitial spaces can also be spaces of 'scape' from the city, spaces of isolation 'within' the city. These meanings speak about the differing perception and functions of interstitial spaces while formally and informally occupied.

The meanings of interstitial spaces also relate to the factors that determined their origins as rural spaces that have been surrounded by the sprawling city or as infrastructural spaces that now connect the outskirts of Santiago with far afield places. These determinants are varied

Figure 7.2 La Platina site. This is an interstitial space that resembles the rural countryside while informally used as landfill.

Source: Author, 2014.

and interlinked and are mainly tied to growth regulations or the absence of regeneration policies. As such, they are placed in a planning system that operates upon individual initiatives on outer lands separated from the peri-urban fringes. Regarding inner interstices, regeneration projects are still embryonic, partially successful, delayed by high costs of infrastructure, or unaffordable for the location of social housing developments. Furthermore, land liberalisation, the inclusion of outer rural villages, and the absence of taxation instruments on empty lands also contribute to land fragmentation and dispersed growth. As excluded from the planning agenda, Santiago's interstitial spaces come to define their own dynamics while becoming marginalised or subject of debates around preservation, integration, and reinforcement of public benefits.

7.3.1 Interstitial spaces as 'built-up' or 'urban' zones

The first determinant of Santiago's interstitiality pertains to the understanding of an 'interstitial space' as such. Although for some actors an interstitial space means 'vacant' or 'undeveloped', for developers in Chile it does not necessarily mean 'empty', 'disintegrated', or 'undeveloped': 'n interstitial space can be a fully urbanised space that does not have enough density in comparison with its surroundings' [interview number 27]. As such, an interstitial space can be a 'built-up' zone – fully integrated into the city's fabric – that has enough capacity to support more density. A low-density can then be an 'interstitial space' in the sense of being lagging behind the densification of its surroundings, a densification that transforms the zone in a spatial 'gap' within a highly densified urban fabric. This is the case of suburban areas in Santiago close to high-quality transport infrastructure but with strong restrictions to densification (Greene et al., 2017), including restrictions related to the preservation of historical heritage in areas where their historical character is debatable [interview number 19; 22]. Similarly, land that can be formally labelled in plans as 'urban' many times is not finally 'urbanised' and thus, remain literally undeveloped for years. This explains why developers argue that there are not clear (or absolute) distinctions between 'empty', 'undeveloped', and 'underused' as categories that define the interstitial condition per se. From a planning perspective, an 'interstitial space' can be a marginalised area irrespective of its degree of emptiness [interview number 4]. This means that the first determinant of interstitiality is not a physical condition but the way of how an interstitial space is conceived within the context of its surrounding urban fabric.

7.3.2 Land atomisation and land privatisation

The 'atomisation of properties' also determines the emergence of interstitial spaces in Santiago. Increasing land subdivision leads to the creation of clusters of small properties that affect large-scale interventions.

When a developer intent to buy a large plot of land to build a large-scale project in an interstitial area composed of clusters of small plots, it becomes complicated as the landowners do not always have interest in selling the land or have differing interests among themselves regarding the future of the area. This is why several areas in Santiago remain as low-density interstitial spaces for years while restraining the construction of metropolitan (private/public) services such as supermarkets, schools, health services, or any other that requires larger land-sizes [interview number 27]. When some landowners of the cluster agree on selling their lands and others do not, the final land-plot geometry is not always ideal for the implementation of simple and functional architectonic layouts, which derives into the creation of pseudo-developed areas composed of a mix of built-up lands interspersed with smaller interstitial spaces. Increasing land privatisation sometimes derives into legal disputes related to heritage and future land uses. While in litigation, land plots remain in stalemate for years and become 'interstitial' [interview number 23]. In some cases, these interstitial spaces are used as temporary parking areas (for a 'meanwhile profit') that makes these interstitial spaces more visible – transitioning from being empty sites into car-parking lots – and also marginal, as they are somehow inactive and unoccupied during the night, weekends or long periods of holiday seasons. The scarcity of larger unified plots, along with the lack of agreement between landowners and litigations, creates interstitial spaces even in areas where development is allowed and services are needed [interview number 19; 23]. These determinants of interstitiality are particularly critical in Chile, given that around 80% of the population have become private homeowners, a direct outcome of the strong promotion of private property as a socially transversal commodity.

7.3.3 Conurbation zones

At a regional scale, Santiago's suburban interstitiality is defined by conurbation zones. Within this context, partial regulations from different institutional frameworks coexist – without equal attributions on land management – and are often defined by differing interests at technical and political levels. These interstitial zones are mainly defined by transport infrastructures and alternations between planned and *de facto* developments that take place almost randomly across the conurbation area [interview number 56]. This process of planned/unplanned urbanisations defines a pseudo-urbanised landscape where different pre-existing land uses coexist with new developments and sub-interstitial spaces of different kinds such as buffers of security, farming lands, industrial areas, geographical restrictions, and others. The rural area between Santiago and Padre Hurtado, for instance, is a recognised conurbation where agricultural activities coexist with railway services, agricultural land-uses, industrial facilities, and housing developments that create tensions regarding the amalgamation of incompatible functions. In that sense, conurbations – as regional spaces – are a context of interstitiality in themselves that can be further examined considering their varied composition and their influence in the urbanisation of surrounding cities.

7.3.4 Enclosing farming and industrial lands

The sprawling expansion of Santiago has spanned previously outer farming and industrial lands that are now within (and part of) the extended suburbia. Somehow, these suburban farming lands are preserved by different factors, including their land uses and legal status as special sites [interview number 31]. A case in point is the area known as 'La Platina' – located in the southern commune of La Pintana – that is labelled as 'rural' considering their functions as an agricultural research centre. As such, the area is under regulations imposed by the Ministry of Agriculture. However, the area is already surrounded by social housing developments and included within the communal urban plan. As 'urban', it is simultaneously subject to regulations related to street maintenance, security, and others that trigger tensions with the MINVU. Also, the local authorities have certain influence on the street maintenance and have included the area within future plans considering its strategic character as an open space for the community. Other cases are some suburban vineyards where wine production is constrained by the surrounding urbanisation. This tension between differing land-uses also extends to industrial lands outside the jurisprudence of the MINVU. This is the case of a chain of extraction sites located in the commune of La Florida. These lands are under the jurisprudence of the Ministry of Mining, but as the pits are immersed within the suburban area, they are also regulated by the Ministry of Housing and local plans defined by the municipality. As such, the area is constrained by a series of ambiguities in governance and urbanisation pressures (Figure 7.3).

There are other cases in a similar situation where 'rurality' and industrial landscapes are immersed within the sprawling expansion of the city. These interstitial spaces create a different category of 'rurality' that is either 'rural' not totally 'urban'; instead, this is 'suburban rurality': 'rural lands engulfed by urban expansion, but that nevertheless still illustrate some degree of farming performance (of any kind) and appear as socio-environmentally active' (Silva, 2020, p. 118). It has been clarified that 'suburban rurality' implies 'a variety of rural lands that benefit the suburbanization process in social, environmental, economic and political terms (...) a category of rurality rarely considered as part of the urban phenomenon, and usually seen as pending space for further (sub)urbanization' (ibid, p. 1). In that sense, Santiago's suburban rurality emerges as a contested field of resistance, non-consensus and political disparity far from being leftovers of the

Figure 7.3 An extraction site in the commune of La Florida.

Source: Author, 2014.

sprawling suburbanisation; interstitial spaces that reclaim their status as social and environmental assets (ibid).

7.3.5 *Administrative boundaries*

Boundary areas between municipalities are critical spaces as they appear as territories of interaction between populations that live in one municipality but work in the other and vice versa. In functional terms, it means that if services belonging to a specific municipality are placed in the municipal boundary area, they also serve the neighbouring population. This defines municipal boundaries as politically ambiguous territories as local mayors prefer to target their interventions at their own constituency, which is to say the population already enrolled as taxpayers and voters within the communal boundary [interview number 33]. Thus, boundary areas become politically abandoned territories, interstitial spaces only of interest to central government that use them for regional infrastructure or for locating metropolitan artifacts such as shopping malls or large-scale facilities that rely on centralised maintenance. This is the case, for instance, of the motorways placed in the boundary area between the communes of Pedro Aguire Cerda and Lo Espejo.

7.3.6 Open tracts as interstitial spaces

Santiago's sprawling growth also describes the presence of abandoned open spaces. This affects private and public lands that become derelict, marginal, and occupied by informal groups. For public authorities, this lack of maintenance relates to the annual evaluation of public expenses that define the base for next year's expenditure [interview number 14]. If empty spaces 'are not part of a politically meaningful project – in the sense of having direct benefits or political credit within the four-year presidential period – they will not be included within the annual investment, and thus, their reconversion and maintenance will be delayed' [interview number 8]. Such spaces include large squares and parks and large inner agricultural sites that are not subject to infrastructural maintenance, security, rubbish removals, street cleaning, and other services. This is particularly ambiguous in large-scale private properties surrounded by low-income neighbourhoods as landowners argue that the surrounding residents informally occupy their sites (throwing rubbish, using the space for illegal activities, etc.), and thus, the land should be maintained using public budget. However, local authorities argue that cleaning, security, and other services cannot be provided for private property. Ultimately, the land enters into an increasingly deteriorating condition that affects the overall quality of the suburban space.

7.3.7 Planning policies and regulations

For some scholars and policy-makers, technical planning instruments derived from policies and regulations are key determinants of the emergence of interstitial spaces in Santiago [interview number 54]. One of these instruments is the 'urban limit' – a containment policy implemented in 1960 through the 'Intercommunal Regulator Plan' (PRI) – that set the city's boundary and restrict urbanisations on rural lands beyond the specified limit. Although designed to control dispersed suburbanisation, the urban limit affects the price of lands that become automatically 'urban' (windfall gain) and encourages landowners to change the function of those portions of land outside the urban limit as their profitability increases with real estate development [interview number 2]. These operations leave empty spaces between the city's boundaries and new developments beyond the urban limit that become interstitial while increasing land fragmentation. Detractors of the 'urban limit' considered its abolition as *panacea* to rectify land-market distortions, increase the land stock, and decrease land prices [interview number 1]. Thus, the 'urban limit' was abolished in 1979 to allow development in any location within the regional space. However, landlords did not sell their lands and kept them undeveloped to catch value over time. Santiago's peri-urban space entered into an unprecedented land-banking exercise that created rings of unaffordable interstitial spaces around the city

(Rodríguez & Sugranyes, 2004). In 1994, this 'urban limit' was finally reinstated, including zoning for future housing developments (MINVU, 1998). Another policy that determines the presence of interstitial spaces is the implementation of 'restriction zones'. These are areas with restricted accessibility that remain undeveloped, underused, or intended for specific and strategic purposes. These are the cases of military sites or industrial lands, ecological reservoirs, and geographical handicaps considered as dangerous for permanent or temporary activities such as water-flooded lands or unstable slopes. In the case of the Puente Alto commune, for example, most of the restriction zones are slopes with 20% of inclination that are inappropriate for urban developments.

7.3.8 Financial constraints

Another factor that determines the presence of interstitial spaces is their financial performance as undeveloped land. Considering that there is no taxation (or impact fees) for keeping a land undeveloped, several empty sites are used for speculation as they will accrue value over time, especially with the arrival of services and infrastructure [interview number 3]. Although the benefits of land speculation are considered to be automatic, these operations perform differently in poor areas as the acquisition of land for services, housing, or infrastructure depends on the purchasing power of the area. Interstitial spaces in poor areas are only attractive for public investments, and it is difficult to keep them well-maintained [interview number 7].

7.3.9 Infrastructural interstitial spaces

Several interstitial spaces in Santiago are remnants of infrastructural services that include motorways, airports, research centres, railway services, military and industrial facilities, farmlands, and water treatment plants. These spaces show decreased levels of functionality although still keep their security buffers or remain underused. These areas cannot be expanded – and thus fall into drabness and disrepair – and are difficult to recover due to the presence of heavy facilities and pollution (Silva, 2017). The communes of Lo Espejo and Pedro Aguirre Cerda, for instance, have inner railway lines and motorways that define large infrastructural spaces immersed within the suburban fabric with impacts at economic, environmental, and political levels [interview number 34]. In Pedro Aguirre Cerda the regional motorways Autopista Central, Autopista del Sol in the north, and Lo Ovalle Avenue in the south, define large interstitial spaces placed within communal boundaries that reinforce spatial segregation at local and metropolitan levels (Figure 7.4). These interstices have a strong impact on residents' daily lives. Further physical barriers – used to improve safety

Figure 7.4 An infrastructural interstitial space at boundary area of Cerrillos and Pedro Aguirre Cerda communes, Santiago de Chile.

Source: Author, 2014.

and pedestrian connectivity – leave them inaccessible and restricted to temporary uses.

Interstitial spaces within Santiago are defined by various determinants possible to be identified as part of the planning process. This means that – while planning mechanisms and policies have historically failed in controlling undesirable urban sprawl – they have simultaneously created distortions that lead to further land fragmentation and thus, the emergence of interstitiality. Although apparently inert, the interstitial spaces of Santiago show different degrees of activity linked to planning rationales aimed to produce urbanised space. The interstitial spaces are elements in which different institutional representations coexist while having various impacts on surroundings that question their marginal condition as invisible or inert spaces excluded from suburban transformations. Although they share a condition of being 'spatial gaps' in the fabric of Santiago in physical and political terms, their diverse nature in terms of origins, functions, and surroundings determine their resistance to standardisation. Indeed, for policy-makers, scholars, residents, and practitioners almost every interstitial space has its own identity and potentials and poses specific challenges that must be addressed case by case.

7.4 Seven cases in Santiago de Chile

The following interstitial spaces were identified by interviewees as relevant and strategic under criteria of location, land capacity – with an
average size of around 300 hectares for each space – and presence of
infrastructure. As such, these criteria point at the opportunities of these
interstitial spaces to serve as catalysers of substantial transformations
in the suburban fabric of Santiago. This sense of relevance partly relates
to the fact that these interstitial spaces are all located in the southern
communes, which present comparatively lower levels of urban quality
in terms of infrastructure, green and public spaces while simultaneously
concentrate high levels of poverty and deprivation. These characteristics
place the southern communes of Santiago as priorities in tackling ine-
quality, distribution of services, territorial disparities, and socio-spatial
segregation (Borsdorf et al., 2016). These interstitial spaces illustrate well
the vicissitudes of Santiago's urban sprawl but mainly the role of interstitial spaces in shaping urban transformations within and beyond the city's
limits.

The order in which the cases are organised describes four types of interstitial spaces. First, two infrastructural sites are analysed to understand how
their relationality connects to the immediate and wider fabric of Santiago
and what are the political perils associated with their reconversion into
city space. One case is a former airport site that is being transformed into
a large residential area equipped with public facilities such as parks and
public spaces. These interstitial spaces can be considered as inner suburban lands. The other case is a military aerial base that is still under operation for minor military practices. The second type of interstitial spaces
relates to rural lands that are still active farming spaces. These interstitial
spaces describe the coexistence between rural and urban functions, the
social and environmental benefits associated with interstices, and the way
of how interstitial spaces define an 'urbanised countryside' in Santiago.
These are the cases of the 'Huertos Obreros y Familiares' [Workers and
Familial Orchards], 'La Platina' site, and 'Campus Antumapu', all located
in the commune of La Pintana. Third, a mining site is analysed to understand the negative effects of interstitial spaces as polluted lands that literally emerge as 'holes' within the sprawling regional fabric. The mining
site is defined by a cluster of gravel pits – all private lands – and as such,
public authorities do not have any incidence on its functions, accessibility,
and plans for transforming the land into housing or any other infrastructure. Like this, there are also other cases in Santiago's sprawling suburbia. Finally, two conurbation zones are analysed in the southern regional
space of Santiago. These conurbations speak about the regional relationality of interstitial spaces while connecting peri-urban areas with regional
functions and hubs such as satellite towns, mining sites, farming lands,
and port-ship zones (Figure 7.5).

[1] Cerrillos Airport site

[2] Military Airbase El Bosque

[3] Campus Antumapu site

[4] La Platina Research Centre

[5] Pits La Florida/Puente Alto

[6] The Orchards of La Pintana

[7] Southern conurbation spaces

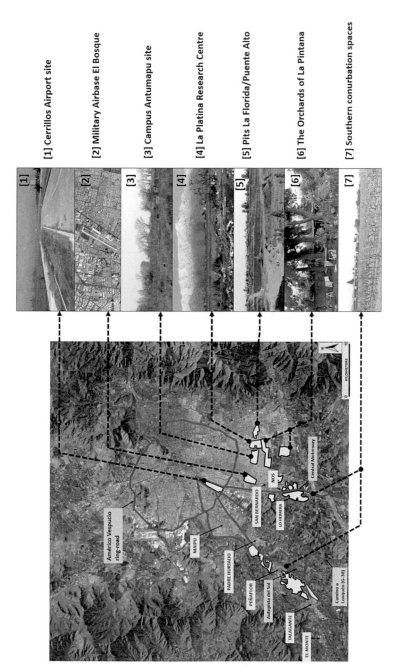

Figure 7.5 Aerial photo of Santiago and the selected interstitial spaces.

Source: Author.

7.4.1 The Cerrillos Airport site

The Cerrillos Airport site is located in the southern municipality of Cerrillos, just at the boundary between the consolidated city and suburban sprawl demarcated by Américo Vespucio ring-road. This site was created in 1929 with a donation from the Guggenheim Foundation to promote aeronautical studies. The site belongs to the Chilean Air Force and was used for military practices. However, in 1967 it became the national airport for civilian flights. Originally its location was in the outer reaches of Santiago and surrounded by open countryside, but with the urban expansion civilian flights were moved into a national airport located outside the city. Then the Cerrillos airport became underused and polluted (including noise, oil, and fuel smell) while simultaneously more valuable in terms of land price.

The factors that made the site visible in the planning agenda were its land capacity (of around 300 hectares), its deprived suburban surroundings, its underused functional condition, its increasing land value, a tense coexistence between aeronautical activities and residential surroundings, and its condition as a spatial and physical barrier. Thus, in 2000 central authorities announced that the area would become the site of 'a new way of making city'; an alternative to the conventional urban sprawl (del Piano, 2010) that would provide high-quality homes, public spaces, parks and would be an open area that would qualify the surrounding suburban space. The reconversion of the area would be a landmark to celebrate the 200 years anniversary of the nation, and thus, the project was named 'Ciudad Parque Bicentenario (CPB)' [Bicentenary Park City]. With an intended completion date of 2010 and fully driven by the MINVU, the project was labelled as pioneer in the implementation of a regeneration policy (Figure 7.6).

By 2001 central authorities made an international call for ideas based on specific planning principles: high construction standards, social integration, environmental sustainability, multi-functional land uses, and flexible urban design as a strategy for adapting the project to market trends (Galilea, 2006). The same year, the MINVU started negotiations with the aerial force, which required another land for military practices. In 2002, the MINVU defined a general Master Plan including the financial model and building codes. The state would make the small-scale urbanisations and a central park to attract real estate investments [interview number 10]. Buildings codes considered four building types to ensure social and spatial diversity [interview number 5]. The proposal also prioritised accessibility, integration, and connectivity [interview number 6]. In 2004, the MINVU closed the airport and social organisations expressed their disagreement while raising legal objections, which left the project in a stalemate for several months (Eliash, 2006). Extra pressure was placed by the Guggenheim Foundation, which argued that the site was to promote aeronautical activities and not urban development. However, the Service of

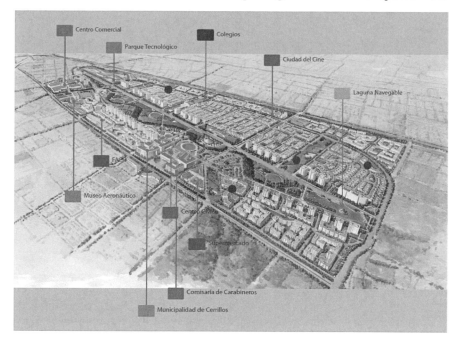

Figure 7.6 Portal Bicentenario project.

Source: MINVU (2013).

Housing and Urbanisation (SERVIU) acquired the land and transformed the area into public property while the demand from the foundation was refused. The unusual predominance of the state, however, triggered a subsequent reaction from developers represented by the Chilean Chamber of Construction (C.Ch.C), which labelled the initiative 'unconstitutional', arguing that the state's role is to deliver subsidies and not urban projects (Instituto Libertad y Desarrollo, 2004). Private firms assumed that the state should transfer the land to them, and then they would provide building models targeted to secure demands. Based on this assumption, all of the urban design proposals converged upon similar socio-economic targets translated into homogeneous architectonic typologies, so 'hard discussions emerged regarding the conception of architecture and urban design as by-products of market niches or as reflections of societal ideals' (Silva, 2017, p. 247). Finally, local authorities rejected proposed densities established in the original proposal, and thus, the project was delayed. By 2011, the only accomplished work was the 'Bicentenary Central Park' – a green area managed by the state and designed by private firms [interview number 9]. By 2014, the landing field remained intact and the park was an isolated desert, contrary to envisaged plans and projected images (Figure 7.7): 'The state is not prepared for this sort of projects. It is as

Figure 7.7 The incomplete Bicentenary Park.

Source: Author, 2014.

simple as that. There is no organic support, neither attributions nor budget. Thus, it depends on the leadership of political and technical teams and our possibilities to convince different actors to align them around common ideas' [interview number 15]. This area is an example of a common issue in Latin American suburban sprawl, where infrastructural services become obsolete and enter into increasing dereliction (Moreira-Arce et al., 2015; Sabatini & Salcedo, 2007). Meanwhile, their reconversion triggers the imaginary of different actors placed at different levels of power, a situation that constantly stimulates the illusion of residents, although the projects are never timely implemented [interview number 25]. These asymmetries immerse the interstitial spaces in endless transitions on their way to become something else.

7.4.2   *The military airbase 'El Bosque'*

The military airbase 'El Bosque' – located in the commune called El Bosque – still maintains minor military practices. This is a spatially closed environment – totally fenced – with restricted public access and totally surrounded by residential areas characterised by low-income families and large social housing developments. This is an almost perfect rectangular site of around 300 hectares that represents 17% of the total communal land, and thus, seen by local authorities as highly relevant to be opened and incorporated into the communal urban fabric. The limits of this site are partly defined by heavy transport infrastructure that contributes to its condition as a closed and isolated environment. In the west – limiting with the commune of San Bernardo – there is a regional railway line that connects Santiago with other cities beyond the metropolitan area. The north and south are the locations of large social housing developments and structural roads. The east side is limited by the structural avenue José Miguel Carrera that connects the whole city with the north–south national highway (Figure 7.8).

This interstitial space is at the heart of one of the most critical communes in terms of poverty concentration, lack of services, and associated social issues of crime, social deprivation, and housing shortages (MDSF, 2019). With a population of 162.505 inhabitants (INE, 2017), the socio-economic profile places the commune among the top 20 districts with high priority for the implementation of social programmes (MDSF, 2019). This is also consistent with the communal average land price of 3.03UF (£91.25)/m^2 in 2021, being one of the cheapest among metropolitan averages (Mercado Inmobiliario, 2021). This is also one of the most densified communes in Santiago with 11.344 people/km^2 (INE, 2017), resulting from a sustained location of social housing developments and detected as a commune that still describes high levels of residential overcrowding [interview number 29]. The communal socio-economic profile has been also historically determined by the relocation of slums and the arrival of relatives of low-income families (MDSF, 2019).

According to local planners, the interstitial condition of the El Bosque site is the typical outcome of an expansive and uncontrolled urban development driven by financial criteria aimed to address housing shortages. This trend has literally infilled all the communal space with low-quality houses and left the airport within the communal fabric. With only a few sites for services, local authorities reclaim the airport for the provision of public amenities such as parks, sports facilities, and others that complement the lack of communal services and workplaces [interview number 29].

Considering that the aerial base is underused and has functionally collided with surrounding residential functions, the municipality has presented several ideas to re-integrate the site into the municipal services. Local authorities and planners face constant negotiations with the MINVU and the Ministry of Defence with no positive results. The arguments raise the advantages for central authorities in recovering the site for centralised goals – such as the accomplishment of 4,500 promised nurseries, housing projects, and sports facilities [interview number 3]. Also, the municipality proposed that only a fraction of 25 hectares can be used for municipal purposes and the rest for the implementation of centralised projects. However, the Air

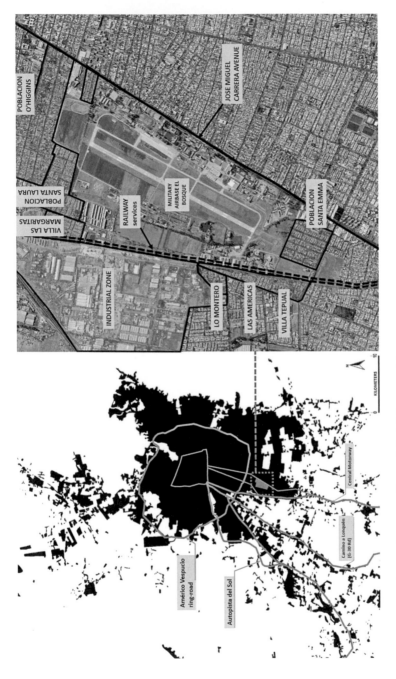

Figure 7.8 The location and surroundings of the El Bosque airbase.

Source: Author.

Force also requires compensation and a portion of land for minor military practices. The final decision relies on central authorities, but the municipality and the community perceive a lack of negotiation capacity and cross-sectorial agreement (MINVU – Ministry of Defence). As a way of gaining visibility in the centralised agenda – and considering that the lack of land for services is a common problem in southern municipalities – an intercommunal municipal association called 'Ciudad Sur' [Southern City] was established to combine political and technical capacities. However, the organisation has different goals and a broad agenda linked to environmental and poverty issues that left the airport site behind other priorities (Asociacion de municipios Ciudad Sur, 2021). Also, political differences among southern municipalities reduced the efficiency of the association linked to the reconversion of underused sites. Additionally, the lack of continuity of military authorities is an extra restriction that impedes the flow of information and the stability of agreements over time [interview number 29].

According to local planners, military authorities are reluctant to transform the site, as it is linked to military history and their sense of identity. This aerial base was the first school of military aviators in Santiago established in 1913 and received the first modern planes for military practices along with the graduation of the first Chilean aviators. The authorities have indeed mounted a small historical museum – hardly open to the public – and have undertaken the construction of new buildings. Some military authorities have shown a certain level of interest to transform the area into a large-scale park – above all because the area is almost underused – but the legal autonomy of the Air Force gives them the chance to manage the future destination of the land with private developers that offer better profitability [interview number 29].

This case illustrates a highly diminished relationality. The area is spatially closed, guarded, and fenced, and its functions define restricted access. Although there is a museum, it is totally marginal considering the land size besides municipal needs. There is an asymmetric scenario of negotiation between the involved actors. The interests of military authorities and the Ministry of Defence are not aligned with the local needs. A low-income district – like El Bosque – has critical needs of services and housing supply and having an interstitial space like the airport site can be the solution. However, the chances to have this site for local purposes would require stronger attention from central agencies and intercommunal arrangements to increase negotiation capacities. This case not only demonstrates that the relational character of an interstitial space can be determined by spatial and functional aspects but also institutional, as military representations restrain any form of integration. This is an interstitial space in which its relevance relates to the characteristics of its surroundings and the opportunities to change a sustained trend of concentrating poverty.

7.4.3 'La Platina' site

'La Platina' is an interstitial space located at the southern commune of La Pintana. This is a 300 hectares site recognised as suitable for large-scale urban projects considering its land capacity, connectivity, and potential impacts at metropolitan levels [interview number 36]. This site is currently well-served by the 'Transantiago' – the modern public transport system implemented in 2007 – that included the improvement of key roads such as the 'Santa Rosa' Avenue (now known as 'Santa Ana Corridor') that connects the whole southern metropolitan area with the city centre. This corridor crosses the entire commune of La Pintana, passing just between the La Platina site and the Campus Antumapu site – another significant interstitial space – to then connect other rural sites and vineyards located at the southern urban fringes. Some suburban mining sites are also connected to this corridor, which finally flows into the 'Acceso Sur' [South Access] corridor that connects the city's core with the main national motorway. So, this is a very well-located interstitial space (Figure 7.9).

'La Platina' currently functions as an agricultural research centre called 'Experimental Station La Platina' under the administration of the 'Instituto Nacional de Investigación Agropecuaria' (INIA) [National Institute of

Figure 7.9 La Platina site at La Pintana commune.

Source: Author, 2014.

Agricultural Research]. This is a public-private corporation originally created with private funds donated by the Rockefeller Foundation and the Chilean Ministry of Agriculture in 1959. The site describes a rural landscape with a few buildings (laboratories and offices). This is mainly an open space used for cultivations and specialised research related to fruits and vegetables, management of plagues and diseases, analysis of quality and resistance of agricultural products, water, and pesticide residues. The centre also hosts the official national bank of vegetal genetic resources [interview number 11]. Although the whole site is intended for agricultural land-uses, the total area for cultivation and research contemplates only 80 hectares (26%). Due to its underused condition, central authorities and the Ministry of Agriculture have proposed a land use change and the relocation of the research centre and its facilities. This change considers 160 hectares for recreational and cultural purposes, 40 hectares for services and housing, and 60 hectares would remain for minor agricultural research activities.

This site is transversally considered as relevant due to its land capacity and its communal context characterised by a high concentration of poverty, lack of services, and a low rate of green spaces regarding minimum standards (CONAMA, 2002). Thereby, the most claimed land use for the site is the creation of an intercommunal park (Ducci, 2002; PLADECO La Pintana, 2012) that includes local, communal, and metropolitan services. These functions are also seen as strategic considering that the site is just in front of another interstitial space of around 320 hectares – the 'Campus Antumapu' – that belongs to the Universidad de Chile that hosts the School of Veterinary, Agronomy and Forest Engineering [interview number 55]. These two sites together define a whole area of 620 hectares, only separated by Santa Rosa Avenue.

This strategic condition and land capacity have been a matter of various initiatives to transform the area into a public space. However, its transformation supposes inter-institutional agreements between different public administrations that at least include the MINVU, the Ministry of Agriculture, the INIA, the Ministry of Public Works, and the municipality of La Pintana. Each of these agencies requires conditions before any agreement; The INIA requires the maintenance of the site, arguing that the location of surrounding social houses has derived into informal occupations and the use of the site as informal landfill. This underused condition is also seen as a social problem as it stimulates the emergence of different manifestations of crime and social misbehaviour. The research institute also indicates that some offenders take the crops and fences illegally and use the site for being hidden during the night. There are also some accusations of harassment to workers and research staff [interview number 11]. The municipality indicates that the site is 'private' (outside its jurisdiction), and thus, it should be cleaned and maintained by the INIA. The MINVU argues that the site is suitable for social housing and services [interview number 3]. However, this idea is rejected by local authorities and residents who see the place as suitable for public parks [interview number 55]. Their idea is to create a green space that transforms the area and increases the overall urban quality.

One of the most ambitious proposals to transform the site is the creation of a metropolitan park (the 'South Park') that joins La Platina, Campus Antumapu, and another adjacent area defined by various gravel pits located at La Florida commune. The proposal contemplates an interconnected park with local, metropolitan, and regional services and includes a new metropolitan Zoo. Studies made by the MINVU (2001–2004) defined the area as suitable for implementing the 'Parque Sur' [South Park], considering a total intercommunal space of around 794 hectares with an impact on about 1,100,000 inhabitants. Financial assessments allowed the provision of green spaces along with sports facilities, lagoons, and the zoo. For the case of La Platina, it would include better quality housing projects with green spaces, and for the Antumapu site it would remain as a large-scale green area. The proposal also included improvements in connectivity (*EMB Construccion Magazine*, 2016) (Figure 7.10).

The project, however, was not implemented, and the area becomes again under discussion for housing developments rather than metropolitan services. Local planners of La Pintana summarise this process as follow:

> We made the study to determine what we can do in these sites. The analysis focused on the social and economic benefits at both communal and intercommunal scales. Then and for different reasons, it was determined that the area must be the place of a large ecological and recreational centre … a mega-park, a park for 'the southern zone' that serves La Pintana and the whole city. After the studies, we realised that central authorities were looking for a place for the metropolitan zoo. So, we applied to locate the zoo in this place. The MINVU agreed, and the project and its implementation were part of a public tender along with the operation of the zoo. The zoo would occupy about 40 hectares of the 162 hectares of the park. However – and after two calls – we did not have successful applications. Then, almost the whole site of La Platina was bought by the SERVIU, which intends using it for social housing. After ten years, we still have nothing and just received a master plan that considers residential uses and only a small green area. We remained very disappointed. The MINVU can fill it with houses if they want, and then we must modify our local plans [interview number 24].

Local residents expect that the place is transformed into 'something else', being the idea of having a park the most accepted one. The views, the green, and the space for community activities provide a hope that the area can be socially safer and that a stronger sense of community can be reinforced [interview number 47]. However, its current condition as a semi-abandoned space is only a matter of informal occupations:

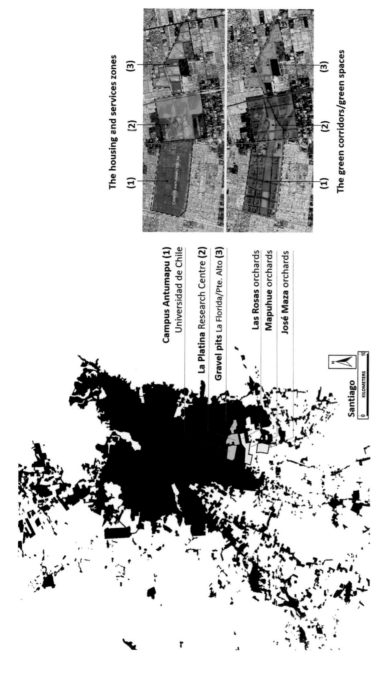

Figure 7.10 The sites of campus Antumapu, La Platina, and the gravel pits joined by the master plan 'Parque Sur'.

Source: *EMB Construccion Magazine* (2016).

If the place remains empty, it would remain as a space for crime and rubbish. We have seen it so many times. So, we would like to have more people who can look after the place ... but not houses. It might be for supermarkets or a police headquarter. As an empty site, the space is suitable for people who want to get drunk ... sometimes some people throw rubbish or break the spotlights to get it dark. It could be a park, but closed and controlled ... otherwise it is so dangerous [interview number 47].

Simultaneously, the site describes some landscape attributes that are transversally seen as positive. Local authorities describe the place as a key environmental asset for the entire metropolitan ecosystem, above all regarding water drainage and air pollution [interview number 37]. Some institutions – such as FAO and environmental NGOs – see the place as a potential 'foodscape' to supply the area with healthy food and encourage the participation of local residents in communal activities [interview number 45]. Local planners and authorities are encouraged to keep the land outside the housing debate and transform this into a large green space:

> ... I understand the site is intended to be densified and I disagree with. I think the role of this space is to offer environmental services for the whole city, which is part of the communal identity. If you consider the thermal layers of Santiago, you will realise that we (the commune) contribute to the reduction of the metropolitan temperature. Why? Because we hold a large stock of agricultural lands. Here, until noon we have fog because of the ground evapotranspiration that remains open. Thus, when the average temperature in Puente Alto is 4°C higher, here is not (...). Also, we have natural water drains through underground aquifers that are some of the largest in the entire region. So, this site must remain as an open space [interview number 37].

Similarly, local residents recognise the relevance of the site as an open space considering its characteristics as a natural landscape. They highlight the beautiful views and the rural aspects as part of the site's aesthetics. These aesthetics also have a meaningful side as residents identify open spaces as components of high-income neighbourhoods usually located closer to the countryside; a characteristic that is given to them by the La Platina site:

> From here it is possible to see the mountains. One can have a look and see the grass that looks beautiful. It was snowing a few days ago and it was just amazing ... like being in the countryside. It is unique. The snow fell over there and you could see everything. The place is good as an open space and must be safe. It would be good if it becomes a park with sport fields ... with many pitches for the whole family [interview number 47].

Figure 7.11 La Platina site. First, it is possible to see the rubbish. Then, some bushes, grass, and trees depict its landscape along with the Andes Mountains at the back.

Source: Author, 2014.

La Platina is an interstitial space of contradictions. The site remains as an open (rural) space with environmental attributes that appeal to the imagination of local authorities and surrounding residents, who argue the site must be preserved [interview number 48; 52]. However, this is also an unsafe space for crime, informal occupations, and pollution. The fact that different institutional representations are operating upon the site suggests further challenges in terms of governance [interview number 12; 13]. In that sense, these interstitial spaces are spaces of convergence of political will and disparities, multiple imaginations, and impacts at multiple scales (Figure 7.11).

7.4.4 'Campus Antumapu'

Campus Antumapu' belongs to the Universidad de Chile. It is a public university, and so, the site is considered as part of the public land stock. This is a 245 hectares site located at La Pintana commune (just in front of La Platina site) currently used by the Faculty of Veterinarian, Agricultural, Forest and Nature Conservation Sciences. The site has administrative offices and student buildings – all part of the architectonic heritage of the city and thus, aimed to be preserved.[1] These constructions and their backyards cover about 60 hectares, which represent 18.75% of the total land. This is mainly an open space for extensive cultivation related to agricultural teaching and research. According to the 'Plan Regulador Metropolitano de Santiago' (PRMS100), this site is an 'Area of Ecological Preservation', and according to the local plan this site is 'excluded from urban development' (Plan Regulador de La Pintana, Memoria Explicativa, 2008). However, both plans allow densification in the boundaries of the site to provide services, sports facilities, and housing. The area remains nevertheless undeveloped and surrounded by social housing neighbourhoods of different types (Figure 7.12). Most of the educational facilities of the university are located nearby Santa Rosa Avenue, which is the main access to the university.

The condition of the site as 'almost empty' is seen by local authorities as negative. This is an underused site while it is well-connected to provide communal and metropolitan services. As an empty space, the site is perceived as a barrier that impedes further connectivity between surrounding neighbourhoods. The university indicates that the site is underused, informally occupied, and unjustified for current educational purposes. However, this condition as an unoccupied site describes its environmental potential as green space and a space that recreates the rural environment found in the outskirts of the city (Figure 7.13).

Despite intensions for changing the land uses, there is still disagreement between the university and the municipality. While the university looks for mechanisms to make the land profitable, local planners need well-located lands for housing projects. Thus, the municipality would allow land-use changes only if the university transfers some hectares for social housing. As there is no agreement, the university made a master plan with new land uses and rented some portions for private agriculture. As the site is labelled 'rural', its land price is still lower than surrounding properties and attractive for private developers. The central government also agrees that the site can be used for social housing and public facilities considering the current low standards of public space and good connectivity in the commune [interview number 18].

Apart from current debates about the future destiny of the site, over the years other projects have been discussed. As previously mentioned, the site was part of the project 'Parque Sur' [South Park] (see Figure 7.11). In this proposal, Campus Antumapu would be the location of new sports facilities, artificial lagoons, housing, and recreational spaces (Universidad de Chile, 2014). The proposal would also include a new stadium for the Universidad de Chile football team.[2] On this basis, various architecture proposals were made considering accessibility, land size, and physical restrictions to support massive encounters. Finally, municipal authorities did not approve the project as a stadium is linked to social misbehaviour; such activities would

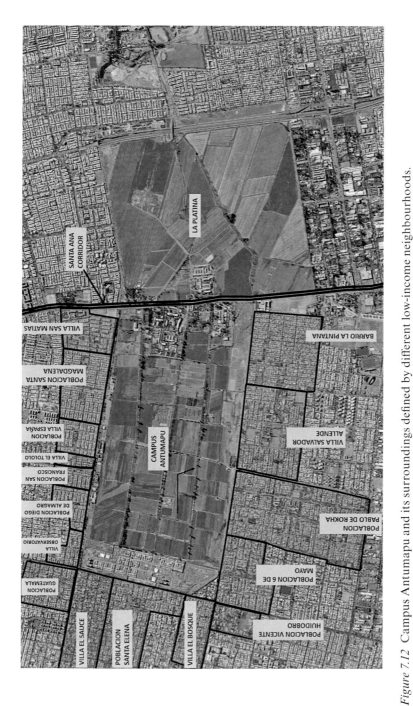

Figure 7.12 Campus Antumapu and its surroundings defined by different low-income neighbourhoods.

Source: Author's map based on the explicative report of the regulator plan of La Pintana, 2008, and google map visited in May 2021.

Figure 7.13 Campus Antumapu and its landscape.

Source: Author, 2016.

trigger negative impacts in a commune mainly formed by low-income population and historically high rates of criminality (Diario AS Chile, 2014;
La Tercera 2020). A final updated master plan – in agreement with the
MINVU – was made to define a new park and residential areas. The new
plan allows communal infrastructure in its boundaries, specifically at
the side of the El Bosque commune. Educational facilities would remain
in the same zone, nearby Santa Rosa Avenue. Green spaces would keep
the rural characteristics of the site. Other areas will be sold to real estate
firms to diversify the social composition, contributing to diminishing the
rate of social segregation. Boundaries made by houses and private properties would define an urban frontage composed of gardens but it would
be also a barrier to avoid undesirable encroachments of the inner space
of the site. The idea is to replace the image of an abandoned land with a
healthy space and configure a relational space that connects with neighbouring communes [interview number 18] (Figure 7.14). This interstitial
space evinces the governance challenges between public institutions – the
MINVU, the municipality, and the university – and how each representation diverts in terms of competing interests. As such, however, there is a
shared understanding about the importance of improving the relationality

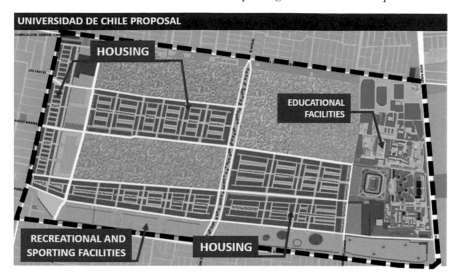

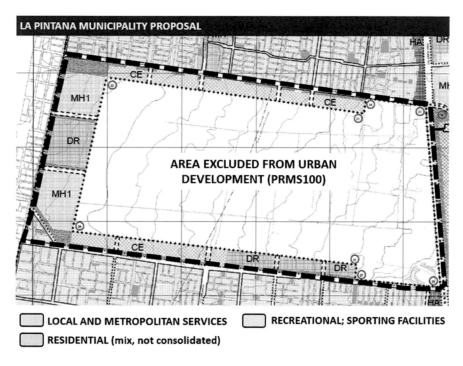

Figure 7.14 At the top, the master plan proposed by Universidad de Chile. Below, the section of the regulator plan of La Pintana.

Source: U. de Chile. Campus Antumapu. Department of infrastructure, 2014; Regulator plan of La Pintana, SECPLA (2008).

of the Campus Antumapu regarding its own connectivity and integration with its surroundings. This site also speaks about the historical significance of interstitial spaces and their potential as part of larger chains of interconnected interstices – La Platina and the gravel pits – that are well-served by metropolitan transport infrastructure.

7.4.5 *Huertos Obreros y Familiares*
[Workers and Familial Orchards]

The case of 'Huertos Obreros y Familiares' [Workers and Familial Orchards] is a highly debated case of social housing policy associated with farming activities. This is an interstitial space located at La Pintana commune that emerged as a cluster of half-hectare land allotments – adjacent to each other – and social housing managed by original families and descendants. As such, this is described as both 'a political and social experiment aimed to create self-sufficient neighbourhoods' [interview number 39]. The orchards were conceived in the 1930s as a social housing model aimed to consolidate productive neighbourhoods, good quality houses, communal land, and local industry (Catalán et al., 2013). The model was inspired by European examples from the nineteenth century (Morán & Aja, 2011; Richter, 2013) and supported by the creation of a housing cooperative (Etxezarreta & Merino, 2013) that coordinated the provision of houses and local agricultural production. This process reached its conclusion in 1941 with the enactment of Law 6.815 – commonly known as the 'José Maza Law' in recognition the Senator José Maza Fernández – which established a permanent fiscal budget to support the agricultural production. In 1943, the government bought a country estate of 315 hectares to allocate 500 land allotments. In 1945, houses were built, and in 1946 the first 150 families arrived to start their lives as urban farmers. In 1951, 156 additional orchards were created. In that period the area was still rural, and families started to ask for water provision, electricity, and sewage. Although the orchards were physically located in the rural space, they were assumed as part of the metropolitan area of Santiago as they were near the urban ratio of commercial feasibility (Roubelat & Armijo, 2012). During the 1970s – and after radical institutional changes dominated by the liberalisation of land and commodification of the social housing supply (Silva, 2020) – the city was rapidly expanded and new social housing typologies reached and surrounded the orchards (Figure 7.15).

  With the centralisation of the housing supply along with the creation of subsidies for low-income families and land privatisation, the expansion of the city towards the south was characterised by the concentration of low-income families. This expansion left the orchard in an interstitial condition while well-served by roads and services. Considering their land capacity and infrastructure, various pressures to buy the orchards emerged with the purpose of building more social housing or leaving the area for services

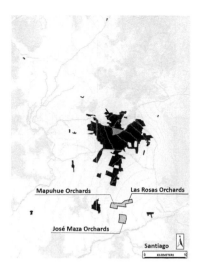
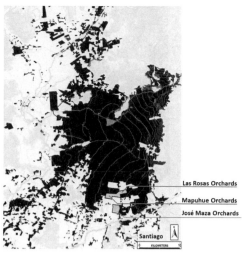

Figure 7.15 The urban expansion of Santiago between 1940 and 2002 and the location of the orchards of La Pintana.

Source: Author's map based on Galetovic & Jordán, 2006.

[interview number 38]. However, families have been resisting urbanisation; they are organised as a community of workers and supported by the still in force 'Jose Maza Law'. Also, some NGOs see the orchards as an example of sustainable urban development and thus, support the families in their resistance against urbanisation [interview number 46]. This resistance emerges from the interstitial condition of the orchards, alternative open spaces that provide food and social stability to the community and the commune of La Pintana.

Over time, the improved reputation of the orchards shifted their perception as poor rural spaces towards spaces of organic food that reach international standards (Gurovich, 2003). The orchards became known for the production of seasonal vegetables, medicinal plants, animal husbandry, seasonal jam and handcrafted meat, vegetable seeds, hydroponics, and organic compost inter alia, all distributed to the rest of the commune, the city, and the region [interview number 40; 41]. Additionally, the orchards are opened to local schools and teachers who bring children to learn about nature, planting, cultivation of local medicines, and healthy nutrition [interview number 42]. This new agenda triggered more initiatives for future projects, strengthening the orchards' resistance to urbanisation:

> If we can expand urban the agriculture projects that we have in the orchards to other places such as parks, public spaces, communal areas, and schools, we would confirm the benefits of the orchards beyond their agricultural functions. The orchards can be another element of an

interconnected network of green spaces of multiple uses without losing their rural identity. The orchards can be a different type of rurality within the city – a green and social infrastructure – that demonstrate that these 'undeveloped' lands are beneficial to the city; not delinked from the city while reducing social marginality [interview number 45].

The orchards of La Pintana are the type of interstitial spaces that speak about the benefits of having the rurality within the city and the resistance to sprawling urbanisation from the social and environmental benefits of suburban rurality. This strong identity has stimulated the organisation of families and neighbours over the years, which have maintained and promoted rurality associated with healthy lifestyles [interview number 43; 44; 48]. Despite the radical institutional changes that occurred in Chile during the 1970s and the rapid urbanisation rate of Santiago, the orchards remain interstitial and perceived as spaces to be preserved. The orchards open further reflections around the social contents of interstitial spaces, preservation and change, and the politics associated with the type of suburbia they suggest (Silva, 2020).

7.4.6 The gravel pits of La Florida/Puente Alto

The gravel pits are an amalgamation of interconnected 'holes' – of around 30 metres depth – resulting from the extraction of raw sand, gravel, and stones used in the construction industry. This is a mining area located just in the administrative boundary between the communes of La Florida and Puente Alto. The mining activities have dragged and excavated the soil for several years, finally creating a three-dimensional landscape that is currently surrounded by low-income neighbourhoods. This interstitial space is close to structural roads and secondary streets that serve the industrial activity and connect the site with regional hubs such as satellite towns and port-ship zones. One of these transport corridors is the Vicuña Mackena Avenue (that further south changes to 'Concha y Toro Avenue') that connects the site with the city centre. This Avenue is part of the Transantiago transport system and delimitates the gravel pits along with the 'Acceso Sur' [South Access] corridor. As mentioned, surroundings are mainly defined by low-income neighbourhoods adjacent to the pits, only separated by local streets and some fences (Figure 7.16).

As a boundary space, the site falls within different municipal arrangements but also central administrations, specifically the ministry of Mining. Also, the MINVU has incidence considering residential surroundings. This is also a system of private properties in the hand of different landowners. Finally, the communal socio-economic and functional profiles of the two communes are different, and thus, they differ in terms of priorities regarding current and future uses for site. La Florida is considered a metropolitan sub-centre (Sánchez et al., 2013), while Puente Alto is one the subsided

Figure 7.16 One of the gravel pits of La Florida/Puente Alto. It is about 30-metres in depth. At the back it is possible to see the residential skyline of two-storey houses.

Source: Author, 2016.

communes due to its low-income socio-economic profile (ibid). Both communes are some of the most populated of Santiago. La Florida has a population of 366,916 inhabitants while Puente Alto has 568,106 inhabitants. Together, they represent 13.14% of the total metropolitan population (INE, 2017). Both communes have also received the largest number of social housing units in the last forty years, representing together the 34.42% of the total stock between 1979 and 2002 (Hidalgo, 2007; Tapia, 2011). All these factors place the gravel pits in a strategic but misaligned position considering their current impacts in terms of pollution and also their future reintegration into the city's fabric.

The gravel pits of La Florida-Puente Alto are literally an open mining site immersed within the suburban fabric of Santiago placed at a communal boundary (Figure 7.17). As a mining area, they are included within the general classification of mining and industrial lands of Santiago, although these activities are located in a residential area. This is a common situation considering that new industries occupy around 968 hectares (46.3%) of restricted land, while around 1,200 hectares in defined zones for exclusive industrial uses remain without occupation (Ducci, 2002). This evinces that

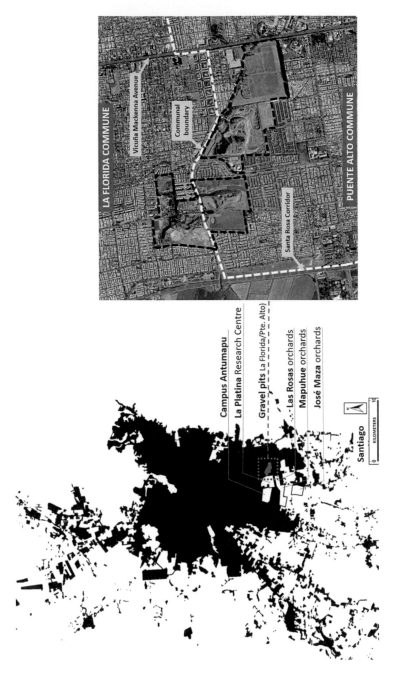

Figure 7.17 At the left, Santiago's map and the location of the gravel pits. At the right, the aerial view of the pits in the boundary between La Florida and Puente Alto.

Source: Author. Image based on google map visited on May 2021.

industrial installations are not determined by the PRMS (MINVUa) but for a case-by-case assessment made by central authorities. Friedmann and Necochea (2014) confirm that the metropolitan region shows 'An extraordinary freedom of localization, perhaps a policy of industrial and residential dispersion and of transport nets and infrastructure driven from central authorities' (p. 75). The gravel pits of La Florida were not originated by Santiago's PRMS plan. Instead, it started first and then became labelled industrial once it became interstitial within Santiago's sprawling fabric [interview number 30].

As the intensity of the industrial activity has decreased over the years, local planners aim to transform the area into a large green space. However, landowners do not see 'a park' as a profitable business in comparison to real estate projects. Furthermore, the landscape is already an uneven geography that requires physical arrangements to become available for any construction or permanent activity, and thus, any transformation would be part of a long-term policy of safety and urban regeneration. Also, there is a lack of coincidence between local and metropolitan plans – that labelled the area as 'green' and 'industrial', respectively. Aside from its boundary location that supposes extra inter-municipal coordination, the two municipalities agree that the area is a 'geographical accident' not allowed for human settlements (PLADECO Puente Alto, 2010). However, this can be transformed into a large green space – part of the 'Green Areas System' project for Santiago – interconnected with other intercommunal parks [interview number 30]. The current development plan of Puente Alto signalises the site as:

> Areas in which there are extractive activities of raw material or abandoned, former landfills and/or affected areas by excavations and/or artificial fillings, established in the article 8.2.1.2 of the PRMS, of soil collapse or settlements identified as zone R5. These zones can be transformed into green areas or sporting and recreational facilities, wells for capturing water.
>
> (Plan Regulador Comunal de Puente Alto, Ordenanza Local, 2002, p. 12)

These debates emerge from the negative connotation of the area in environmental, social, and economic terms. At present. it appears as a closed and unsafe environment without light during the night, and its boundaries are informally occupied and used as landfills. The industrial activity is a factor of air pollution due to soil movements and circulation of trucks, which also impacts streets maintenance. The surrounding population permanently complains about 'suspended dust' that affects people's health.

The gravel pits of La Florida/Puente Alto speak about the relationality of interstitial spaces and their condition as boundary objects. While the site is closed to its immediate surroundings, it connects to wider networks of economic activity linked to the construction industry and other regional functions. As a boundary object, it is also the space of convergence of different public-private actors and different levels of governance. Its literal location in the boundary space between the two municipalities confirms that interstitial spaces can operate as elements of intersection of interests and representations, while they nevertheless remain interstitial and in a stalemate for years. The gravel pits are one example of many sites that originally emerged as interstitial hubs located in the regional space beyond the city's limits. However, with the arrival of Santiago's sprawling suburbia it became interstitial, a process of change in which interstitial hubs become incompatible elements with the city's functions and, thus, spaces to be expelled from the city's hinterland.

7.4.7 The southern conurbations

The southern conurbations of Santiago define another category of interstitiality that illustrates different relationalities and morphologies. They are formed by different elements that include interstitial hubs and sub-interstitial spaces interspersed with built-up areas of different sorts. Conurbations are a complex mix of unregulated and regulated lands that define an ambiguous landscape in which residential, farming areas, and infrastructures coexist. Owing to the infrastructures and lack of constraints on rural lands, the resulting interstitial spaces become accessible to new developments with impacts at economic and environmental levels. Cases in point include the conurbations between Maipú and Padre Hurtado – structured by the 'Autopista del Sol' [The Sun Motorway] and 'Camino a Melipilla' [Melipilla Road] – and the conurbation between southern Santiago and San Bernardo structured by the 'Autopista Central' [Central Motorway] (Figure 7.18).

The conurbation between Maipu and Padre Hurtado is primarily explicated by the growth experienced by the commune of Maipú – that along with Puente Alto – has concentrated the 70% of the increment of population after 1978, being the two most populated communes in Chile between 1982 and 2002 (Tokman, 2006). Empirical studies show a clear concentration of social housing units between 1979 and 2002 (Hidalgo, 2007; Tapia, 2011) and an increment in communal land prices (Brain & Sabatini, 2006). Therefore, the communes were included in the last modification of the PRMS100. In 1994, the urbanised area of Maipú reached the boundary of the neighbouring commune of Padre Hurtado, which was also included in the PRMS100. By diminishing the hectares of rural land, this rapid growth has been described as an environmental hazard for Santiago de Chile (Krellenberg et al., 2013; Romero & Órdenes, 2004). This urbanisation process has also been driven by the improvement of the transport infrastructure in which the 'Autopista del Sol' [the Sun Motorway], and

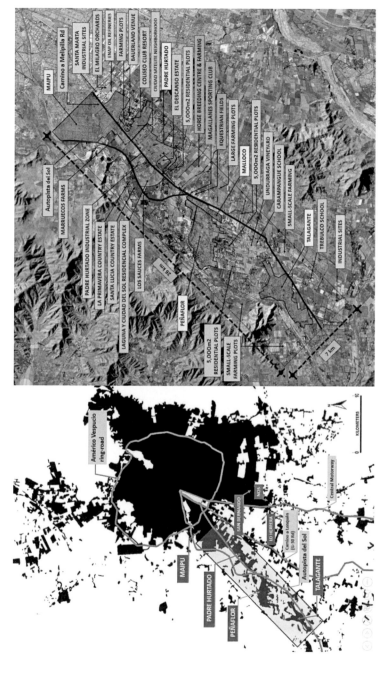

Figure 7.18 Santiago's southern conurbations.

Source: Author.

'Camino San Alberto Hurtado' [St Alberto Hurtado Road] connect Maipú with Padre Hurtado, Peñaflor, Talagante, El Monte, and finally Melipilla. The latter is a former rural village that has experienced an exponential growth due to the massive location of social housing developments and rural-urban migrations (INE, 2017). This is a conurbation process that involves various local municipalities that constantly update their local plans to incorporate new social housing projects and infrastructure delivered by the central government. These policies are seen by local authorities as a concrete threat to local planning decisions that would affect both local and metropolitan functions alike:

> Central authorities decide the location of new social housing and market housing in our communal land. However, they do not provide proportionate responses in transport infrastructure. That's why we would need to be compensated for the externalities of these decisions. As local authorities, we are responsible of the street maintenance, future extraction of domestic rubbish and the maintenance of green spaces. More importantly, we will deal with the collapse of our structural connectivity as the new plot subdivisions means more houses and so, more cars. We will see problems of transport congestion, etc – i.e. if either the government nor the real estate firms provide the roads, how can we deal with these problems in the future? They only considered the access to the 'Autopista del Sol', completed just in 2013. The central authorities are in debt with the commune in terms of connectivity … i.e. how do they get the road to Melipilla without having impacts in our commune? We are forced, however, to approve all real estate projects and all of them will get the road to Melipilla … from here and there … without a proper capacity. This road is not properly designed for this demand; this is a rural road [interview number 35].

Another case in point is the conurbation between San Bernardo and Lo Herrera. This conurbation process occurs within the same municipal jurisprudence (San Bernardo). The commune of San Bernardo is also included within PRMS100, a modification that increased the communal available land from 5,400 hectares to almost 9,000 hectares (more than 70%), in which 64% is destined for residential uses (Boccardo, 2011). San Bernardo has also changed its urban limits to include more housing developments, green spaces, connectivity, and to reconvert industrial lands. The commune is considered a 'sub-centre' of Santiago, and an independent area with high levels of functional self-sufficiency (PLADECO San Bernardo, 2011). The plan includes further densification of a small rural village called 'Lo Herrera' and other rural villages intended to support future growth. Some of the lands located in these villages are labelled as 'special uses', others as buffers of security, underused facilities, military installations, and sports facilities. These lands are interspersed with a range of rural plots also intended to be urbanised. Some inner interstitial spaces – such as the Cerro Chena and a large gravel pit – appear as geographical restrictions; others as abandoned spaces that 'can be part of the environmental assets of the commune if properly included in our plans for green space' [interview number 26].

This conurbation process takes place within the municipal boundaries. However, the conurbation is still triggered by centralised policies in social housing and transport infrastructure. In that sense, the local governance does not make a substantial difference regarding land management of the conurbation while the commune indeed updates its local plans accordingly. This is why the San Bernardo/Lo Herrera's conurbation describes a mix of planned and unplanned spaces – many of them in an interstitial condition – that vanishes the boundaries between the urban and the rural, with impacts in the landscape quality of the conurbation space and its associated urbanised areas.

As seen, both conurbations reach a regional scale of influence in terms of magnitude and relationality. These conurbations involve different districts, rural lands, industrial areas, and infrastructures. As such, they are multifaceted in terms of governance and spatial composition. The centralised actions over the conurbation space are not coordinated with local actors – neither between intercommunal authorities of Maipu and Padre Hurtado, for instance, or the intra-communal authorities of San Bernardo – which poses current and future challenges in urban governance. At present, these interstitial spaces appear as somehow passive receptors of centralised and private operations over the rural space (Figure 7.19).

7.5 Conclusions

The interstitial spaces of southern Santiago are relevant components of the sprawling geography of the city and are placed at the core of the planning processes that define Santiago's urban development. While different planning determinants explain the emergence of interstitial spaces, it is clear that planning policies play an important role in land fragmentation and the meanings of interstitial spaces as part of. Including less densified zones seen as gaps within the urban fabric, atomised lands that remain undeveloped, encroached rural lands, administrative boundaries between municipalities, open tracts of different sorts, lands used for land banking, and infrastructural lands, the interstitial spaces of Santiago confirm their context dependant character and their importance as elements produced alongside the urbanisation process.

The analysed cases demonstrate that interstitial spaces are multifaceted elements in terms of functions, morphology, scale, relationality, and also the governance implicit in the interaction of multiple institutional representations linked to their current and future land uses. In that sense, the interstitial spaces pose various challenges in governance and call for

Figure 7.19 An interstitial space in the conurbation zone of Maipu/Padre Hurtado.

Source: Author, June 2016.

interdisciplinary approaches in the understanding of their infrastructural, social, environmental, economic, functional, and political values. The interstitial spaces of Santiago are also relevant for their surroundings. Mostly immersed in low-income areas, their land capacity, infrastructures, and locations trigger the imagination of different actors regarding their current environmental values and future land uses. However, different actors raise different views that somehow leave the transformation of Santiago's interstitial spaces in a stalemate or in the hands of centralised decision making. What is clear is that their current conditions are somehow ambivalent: while for some they represent freedom and proximity to nature, for others they are derelict, abandoned, and marginalised spaces.

The interstitial spaces of Santiago's urban sprawl are not empty lands. They describe multiple functions, impacts, values, and potentials that influence the suburban transformation of the city. At least, the interstitial spaces of Santiago are instances of socio-political illusion to change urbanisation trends and improve (sub)urban standards; as such, they disclose the multiple political perils placed at metropolitan and regional scales to manage their integration.

Notes

1 Buildings of the Campus are recognised by their architectonic value. They are part of the Chilean 'architectonic modern heritage' built at the end of the 1960s by reputed architects that followed the Bauhaus tradition. In this case, the design of the buildings was directly supported by a former member of the Bauhaus, the architect Tibor Weiner (U. de Chile, Master Plan Campus Antumapu, April 2014).
2 The football club 'Universidad de Chile' is one of the top professional football clubs in Chile. Founded in 1927 has historically pursued the creation of its own stadium (Club Universidad de Chile: https://www.udechile.cl/).

References

Asociacion de municipios Ciudad Sur (2021): http://www.municipiosciudadsur.cl/portal/ejecutados/

Barton, J., Román, Á., & Fløysand, A. (2012). Resource Extraction and local justice in Chile: Conflicts over the commodification of spaces and the sustainable development of places. In H. Haarstad (Ed.), *New political spaces in Latin American natural resource governance* (pp. 107–128). Palgrave Macmillan.

Boccardo, D. (2011). Tensions of a triple urban vocation: San Bernardo and its absortion process from Santiago de Chile. *Revista Terrtitorios en Formación*, 1(2), 7–20.

Borsdorf, A., & Hidalgo, R. (2010). From polarization to fragmentation. Recent changes in Latin American urbanization. In P. Lindert, O. Verkoren (Eds.), *Decentralized development in Latin America* (pp. 23–34). Springer.

Borsdorf, A., Hidalgo, R., & Sánchez, R. (2007). A new model of urban development in Latin America: The gated communities and fenced cities in the metropolitan areas of Santiago de Chile and Valparaíso. *Cities*, 24(5), 365–378.

Borsdorf, A., Hildalgo, R., & Vidal-Koppmann, S. (2016). Social segregation and gated communities in Santiago de Chile and Buenos Aires. A comparison. *Habitat International*, 54, 18–27.

Brain, I., & Sabatini, F. (2006). Los precios del suelo en alza carcomen el subsidio habitacional, contribuyendo al deterioro en la calidad y localización de la vivienda social (The increasing land prices erode the housing subsidy and deteriorate the quality and location of the estate housing). *Revista Pro Urbana*, 4, 2–13.

Cámara Chilena de la Construcción, C. C. C. (2012). Disponibilidad de suelo en el gran Santiago (Land availability in Greater Santiago) Resultados Estudio 2012, Evolución 2007–2012. Gerencia de Estudiosm C.Ch.C., Santiago de Chile.

Ministerio de Desarrollo Social y Familia (MDSF) de Chile (2019). Encuesta de caracterización socioeconómica, CASEN: https://www.desarrollosocialyfamilia.gob.cl/storage/docs/Informe_de_Desarrollo_Social_2019.pdf

Catalán, J., Fernandez, J., & Olea, J. (2013). *Cultivando Historia. Trayectorias, Problemáticas y Proyecciones de los Huertos de La Pintana*. Dhiyo.

Clichevsky, N. (2007). La tierra vacante 'revisitada': elementos explicativos y potencialidades de utilización (Vacant lands 'revisited': Explicative elements and land-use potentials). *Cuaderno urbano: espacio, cultura y sociedad*, 1(6), 195–219.

Club Universidad de Chile (2021). https://www.udechile.cl/ (accessed in May 2021).

Cox, T., & Hurtubia, R. (2021). Subdividing the sprawl: Endogenous segmentation of housing submarkets in expansion areas of Santiago, Chile. *Environment and Planning B: Urban Analytics and City Science*, 48(7), 1770–1786.

de Mattos, C. (1999). Santiago de Chile, globalización y expansión metropolitana: Lo que existía sigue existiendo. *EURE, 225*(76), 29–56.

de Mattos, C. (2001). Metropolización y suburbanización. *EURE, 27*(80), 5–8.

de Mattos, C., Fuentes, L., & Link, F. (2014). Tendencias recientes del crecimiento metropolitano en Santiago de Chile. ¿Hacia una nueva geografía urbana? (Recent metropolitan growth trends in Santiago de Chile. Towards a new urban geography?). *INVI, 29*, 193–219.

del Piano, A. (2010). *Memoria explicativa, Ciudad Parque Bicentenario – Una nueva forma de hacer ciudad (Executive report Bicentenary Park City – A new way of making city)*. Ministerio de Vivienda y Urbanismo (MINVU), Chile.

Diario AS Chile. (2014). Los detalles del futuro estadio de la U. de Chile en La Pintana (The details of the new stadium of the U. de Chile at La Pintana). https://chile.as.com/chile/2014/12/13/futbol/1418488129_715953.html (accessed in May 2021).

Dubeaux, S., & Sabot, E. C. (2018). Maximizing the potential of vacant spaces within shrinking cities, a German approach. *Cities, 75*, 6–11.

Ducci, M. (2002). Área urbana de Santiago 1991–2000: Expansión de la industria y la vivienda. *EURE, 28*(85), 187–207.

Ducci, M., & Gonzalez, M. (2006). Anatomía de la expansión de Santiago, 1991–2000 (Anatomy of Santiago's expansion, 1991–2000). In A. Galetovic (Ed.), *Santiago: Dónde estamos y hacia dónde vamos* [*Santiago: where do we are and where do we go*] (pp. 123–46). Centro de Estudios Públicos.

Eliash, H. (2006). Portal Bicentenario: breve crónica de un Proyecto emblemático (Bicentenary portal: A short chronicle of a flagship project). *Revista de Arquitectura, Universidad de Chile, 13*, 56–58.

EMB Construccion Magazine. (September 2016).Interview with Arturo Lyon. Proyecto Sur: Una respuesta diferencte frente al crecimiento urbano (South Project: A different response to face the urban growth). http://www.emb.cl/construccion/articulo.mvc?xid=1814&edi=81&tip=&act=&srch=&rand=4216X2229&cmtok=x#cmt (accessed in May, 2021).

Etxezarreta, A., & Merino, S. (2013). Las cooperativas de vivienda como alternativa al problema de la vivienda en la actual crisis económica (The housing cooperatives as alternative to the housing problem in the current economic crisis). *REVESCO. Revista de Estudios Cooperativos, 113*, 92–119.

Flores, M., Otazo-Sánchez, E. M., Galeana-Pizana, M., Roldán-Cruz, E. I., Razo-Zárate, R., González-Ramírez, C. A., & Gordillo-Martínez, A. J. (2017). Urban driving forces and megacity expansion threats. Study case in the Mexico City periphery. *Habitat International, 64*, 109–122.

Friedmann, J., & Necochea, A. (2014). Algunos problemas de política de urbanización de la Región Capital de Chile (Some issues of the urbanisation policy in the capital region of Chile). *EURE, 1*(1), 63–95.

Fuentes, L., & Pezoa, M. (2018). Nuevas geografías urbanas en Santiago de Chile 1992–2012. Entre la explosión y la implosión de lo metropolitano (New urban geographies in Santiago de Chile 1992–2012. Between explosion and implosion of the metropolitan). *Revista de Geografía Norte Grande*, (70), 131–151.

Gainza, X., & Livert, F. (2013). Urban form and the environmental impact of commuting in a segregated city, Santiago de Chile. *Environment and Planning B: Planning and Design, 40*(3), 507–522.

Galetovic, A., & Jordán, P. (2006). Santiago: ¿dónde estamos?, ¿hacia dónde vamos? [Santiago: Where do we are?, where do we go?] In A. Galetovic, Santiago: dónde estamos y hacia dónde vamos [Santiago: where we are and were we go]. (pp. 25–69). CENTRO DE ESTUDIOS PÚBLICOS.

Galilea, S. (2006). Proyecto Portal Bicentenario (Bicentenary Portal Project). *Revista de Arquitectura*, *13*, 58–61.

Gavrilidis, A. A., Niță, M. R., Onose, D. A., Badiu, D. L., & Năstase, I. I. (2019). Methodological framework for urban sprawl control through sustainable planning of urban green infrastructure. *Ecological Indicators*, *96*, 67–78.

Greene, M., Mora, R. I., Figueroa, C., Waintrub, N., & Ortúzar, J. D. D. (2017). Towards a sustainable city: Applying urban renewal incentives according to the social and urban characteristics of the area. *Habitat International*, *68*, 15–23.

Gross, P. (1991). Santiago de Chile (1925–1990): Planificación urbana y modelos políticos (Santiago de Chile (1925–1990): Urban planning and political models). *EURE*, *17*(52/53), 27–52.

Gurovich, A. (2003). Conjugando los tiempos del verbo idealizar: los huertos obreros y familiares de La Pintana, Santiago de Chile (Conjugating the verb idealising: the workers and familial orchards of La Pintana, Santiago de Chile). *Cuadernos del CENDES*, *20*(53), 65–76.

Heinrichs, D., & Nuissl, H. (2015). Suburbanisation in Latin America: Towards new authoritarian modes of governance at the urban margin. In P. Hamel, & R. Keil (Eds.), *Suburban governance. A global view* (pp. 216–238). University of Toronto Press.

Heinrichs, D., Nuissl, H., & Rodriguez, C. (2009). Dispersión urbana y nuevos desafíos para la gobernanza (metropolitana) en América Latina: el caso de Santiago de Chile. *EURE*, *35*(104), 29–46.

Heinrichs, D., Lukas, M., & Nuissl, H. (2011). Privatization of the fringes – A Latin American version of post-suburbia? The case of Santiago de Chile. In *International perspectives on suburbanization* (pp. 101–121). Palgrave Macmillan.

Herzog, L. (2015). *Global suburbs. Urban Sprawl in Rio Grande to Rio de Janeiro*. Routledge.

Hidalgo, R. (2007). ¿Se acabó el suelo en la gran ciudad? Las nuevas periferias metro-politanas de la vivienda social en Santiago de Chile (Is the land over in the greater city? The new metropolitan peripheries of state housing in Santiago de Chile). *EURE*, *33*(98), 57–75.

Instituto Libertad y Desarrollo. (2004). ¿Estado Empresario? Casos Cerrillos y ENAP Perú (Entrepreneurial state? The case of Cerrillos and ENAP Perú). In *Temas Publicos*, *N°689*. Instituto Libertad y Desarrollo.

Instituto Nacional de Estadisticas (INE). (2017). *Censo de Poblacion y Vivienda* ((National Institute of Statistics, INE. *Statistics of population and housing*). https://www.ine.cl/estadisticas/sociales/censos-de-poblacion-y-vivienda/poblacion-y-vivienda (accessed in March 2021).

Krellenberg, K., Müller, A., Schwarz, A., Höfer, R., & Welz, J. (2013). Flood and heat hazards in the Metropolitan Region of Santiago de Chile and the socio-economics of exposure. *Applied Geography*, *38*, 86–95.

La Tercera. (2020). Un sueño constantemente frustrado: los numerosos intentos de la U por tener un estadio (A constantly frustrated dream: the countless attempts of the Universidad de Chile club (U) for having a new stadium). Available (in Spanish) at:

https://www.latercera.com/el-deportivo/noticia/un-sueno-constantemente-frustrado-las-numerosos-intentos-de-la-u-por-tener-un-estadio/Y24BTQKU3 NBKBCT2BQ72UMWR6A/ (accessed in May 2021).

Link, F. (2008). From polycentricity to fragmentation in Santiago de Chile. *Centro-h, Revista de la Organización Latinoamericana y del Caribe de Centros Históricos*, *2*(02), 13–24.

Mercado Inmobiliario. (2021). http://mercadoinmobiliarioper.blogspot.com/2013/06/valor-del-suelo-por-comuna.html

Ministerio de Vivienda y Urbanismo, MINVU. (1998). Memoria explicativa del Plan Regulador de Santiago. Incorporación de las comunas de Colina, Lampa y Til-Til. Santiago, Chile (Executive report of the Development Plan of Santiago. Incorporation of the Colina, Lampa y Til-Til communes).

Ministerio de Vivienda y Urbanismo, MINVU. (2014a). Texto Actualizado y Compaginado, Ordenanza Plan Regulador Metropolitano de Santiago (PRMS) (Updated and compiled text of the Building Codes, Metropolitan Development Plan of Santiago). http://metropolitana.minvu.cl/pag-m/documentacion-vigente-prms/ (accessed in March 2021).

Ministerio de Vivienda y Urbanismo, MINVU. (2014b). *Towards a New Urban Policy for Chile: National Policy of Urban Development*. https://cndu.gob.cl/wp-content/uploads/2014/10/L4-Politica-Nacional-Urbana.pdf (accessed in March 2021).

Ministerio del Medio Ambiente. Corporación Nacional de Medio Ambiente, CONAMA. (2002). *Áreas verdes en el Gran Santiago (Green áreas in Greater Santiago)*. Área de Ordenamiento Territorial y Recursos Naturales. Conama, Santiago de Chile.

Morán, N., & Aja, A. (2011). *Historia de los huertos urbanos. De los huertos para pobres a los programas de agricultura urbana ecológica (History of the urban orchards. From the orchards for the poor to the ecological urban agricultura programmes)*. Departamento de Urbanística y Ordenación del Territorio, Escuela Técnica Superior de Arquitectura de Madrid (UPM).

Moreira-Arce, D., De la Barrera, F., & Bustamante, R. O. (2015). Distance to suburban/wildland border interacts with habitat type for structuring exotic plant communities in a natural area surrounding a metropolitan area in central Chile. *Plant Ecology & Diversity*, 8(3), 363–370.

Phelps, N. A., & Silva, C. (2017). Mind the gaps! A research agenda for urban interstices. *Urban Studies*, 55(6), 1203–1222.

Plan de Desarrollo Comunal de La Pintana, PLADECO 2012–2016. (2012). [Communal Development Plan of La Pintana]. Secretaria de Planificación Comunal. I. Municipalidad de la Pintana. Santiago, Chile.

Plan de desarrollo Comunal de Puente Alto, PLADECO (2010) [Communal Development Plan of Puente Alto]. I. Municipalidad de Puente Alto, Puente Alto. Santiago, Chile.

Plan de Desarrollo Comunal de San Bernardo, PLADECO (2011) [Communal Development Plan of San Bernardo]. Secretaria Comunal de Planificación, SECPLA. I. Municipalidad de San Bernardo. Santiago, Chile.

Plan Regulador Comunal de Puente Alto (2002). Ordenanza Local. Secretaria de Planificación Comunal, SECPLA [Communal Development of Puente Alto: Local Building Codes]. I. Municipalidad de Puente Alto. Santiago, Chile.

Plan Regulador de La Pintana. Memoria Explicativa (2008) [Executive Report of the Local Development Plan]. I. Municipalidad de La Pintana. Secretaria Metropolitana de Vivienda y Urbanismo. Santiago, Chile.

Richter, F. (2013). La agricultura urbana y el cultivo de sí. Los huertos de ocio a la luz de las dinámicas neorrurales (Urban agriculture and cultivation. The leisure orchards under neorural dynamics). *ENCRUCIJADAS. Revista crítica de Ciencias Sociales, 6,* 129–145.

Rodríguez, A., & Sugranyes, A. (2004). El problema de vivienda de los 'con techo' (The housing problem of those 'with a roof'). *EURE, 30*(91), 53–65.

Rojas, C., Muñiz, I., & Pino, J. (2013). Understanding the urban sprawl in the mid-size Latin American cities through the urban form: Analysis of the conception metropolitan area (Chile). *Journal of Geographic Information System,* 5(3), 222–234.

Romero, H., & Órdenes, F. (2004). Emerging urbanization in the Southern Andes. Environmental impacts of urban sprawl in Santiago de Chile on the Andes Piedmont. *Mountain Research and Development, 24*(3), 197–201.

Romero, H., Vásquez, A., Fuentes, C., Salgado, M., Schmidt, A., & Banzhaf, E. (2012). Assessing urban environmental segregation (UES). The case of Santiago de Chile. *Ecological Indicators, 23,* 76–87.

Roubelat, L., & Armijo, G. (2012). Urban agriculture in the metropolitan area of Santiago de Chile. An environmental instrument to create a sustainable model. *PLEA 2012 – 28th Conference, Opportunities, Limits & Needs Towards an Environmentally Responsible Architecture Lima,* Perú 7–9 November 2012.

Sabatini, F. (2015). Hacia una Politica de integracion social urbana: cinco carencias de la Politica Nacional de Desarrollo Urbano [Towards a policy of urban social integration: five drawbacks of the National Policy of Urban Development]. In V. Paz-Mellado (Ed.), *La ciudad que queremos [The city we want]* (pp. 63–84). Biblioteca del Congreso Nacional. Departamento de Estudios, Extension y Publicaciones.

Sabatini, F., & Salcedo, R. (2007). Gated communities and the poor in Santiago, Chile: Functional and symbolic integration in a context of aggressive capitalist colonization of lower-class areas. *Housing Policy Debate, 18,* 577–606.

Sabatini, F., Wormald, G., Sierralta, C., & Peters, P. A. (2009). Residential segregation in Santiago: Scale-related effects and trends, 1992–2002. In B. R. Roberts, & R. H. Wilson (Eds.), *Urban segregation and governance in the Americas* (pp. 121–143). Palgrave Macmillan.

Sánchez, R., Velasco, F., & Medina, M. (2013). Modelación espacial para la identificación de subcentros de empleo en el Gran Santiago (Spatial modelling for identifying employment subcentres in Greater Santiago). In Universidad de Chile (Ed.), *XIV Congreso Chileno de Ingeniería de Transporte* (No. 16) (pp. 1–15). Santiago, Chile.

Sieverts, T. (2003). *Cities without cities. An interpretation of the Zwischenstadt.* Spon Press.

Silva, C. (2017). The infrastructural lands of urban sprawl: Planning potentials and political perils. *Town Planning Review, 88*(2), 233–256.

Silva, C. (2019). The interstitial spaces of urban sprawl: Unpacking the marginal suburban geography of Santiago de Chile. In N. Geraghty, & A. Massidda (Eds.), *Creative spaces: Urban culture and marginality in Latin America* (pp. 55–84). University of London Press. https://humanities-digital-library.org/index.php/hdl/catalog/book/creative_spaces

Silva, C. (2020). The rural lands of urban sprawl: Institutional changes and suburban rurality in Santiago de Chile. *Asian Geographer, 37*(2), 117–144.

Silva, C., & Vergara, F. (2021). Determinants of urban sprawl in Latin America: Evidence from Santiago de Chile. *SN Social Sciences, 1*(202), 1-35.

Tapia, R. (2011). Social housing in Santiago. Analysis of its locational behavior between 1980-2002. *INVI, 73*(26), 105–131.

Tokman, A. (2006). El MINVU, la política habitacional y la expansión excesiva de Santiago (The MINVU, the housing policy, and the excessive expansion of Santiago). In A. Galetovic (Ed.), *Santiago: Dónde estamos y hacia dónde vamos* (pp. 489–520). Centro de Estudios Públicos.

Trivelli, P. (2015). Algunas reflexiones sobre la Política Nacional de Desarrollo Urbano, la Economía Urbana y el financiamiento de las ciudades (Some reflections about the National Policy of Urban Development, the urban economy and financing of cities). In *La ciudad que queremos (The city we want)* (pp. 85–108). Biblioteca del Congreso Nacional. Departamento de Estudios, Extension y Publicaciones.

Truffello, R., & Hidalgo, R. (2015). Policentrismo en el Área Metropolitana de Santiago de Chile: Reestructuración comercial, movilidad y tipificación de subcentros (Polycentrism in the metropolian área of Santiago de Chile: comercial restructuring, mobility and the categorisation fo subcentres). *EURE, 41*(122), 49–73.

UN. (2018). *The World's Cities in 2018*. https://population.un.org/wup/Publications/ (accessed in April 2021).

Universidad de Chile. (April 2014). Campus Antumapu. Presentación del Plan Maestro. Universidad de Chile. Campus Sur (Antumapu). http://www.uchile.cl/portal/presentacion/campus/7987/campus-sur (accessed in May 2021).

Vergara, F., & Boano, C. (2020). Exploring the contradiction in the ethos of urban practitioners under neoliberalism: A case study of housing production in Chile. *Journal of Planning Education and Research*, 1–15.

8 The implications of interstitial spaces in urban studies

> where there is nothing, everything is possible; where there is architecture, nothing (else) is possible.
>
> (Koolhaas, 1995, p. 199)

8.1 Introduction

It has been clarified that interstitial spaces are the spatial gaps that lie within, between, and beyond built-up lands and cities – built-up lands that have distinctive characteristics of 'place' of some kind or another – and where different kinds of economic, social, or political activity in the forms of interconnected hubs can (or cannot) manifest. The interstitial spaces are 'interstitial' in the sense of being in-between entities, and 'spaces' in the sense of being opposite to those areas that can properly be identified as 'places'; in the sense of embracing Lefevbre's abstract space that 'is inherently violent and geographically expansive' (Brenner & Elden, 2009, p. 359). As such, the interstitial spaces emerge at different geographical scales – proximity, transition, regional, and remoteness – and describe different degrees of relationality while connecting or isolating surroundings. Considering their geographical scales, the interstitial spaces can be seen within, between, or beyond cities and regions as they irrigate as liminal spaces at the very core of cities (Arvanitis et al., 2019) up to the remote open countryside. This system of interstitial spaces configures the 'interstitial geography' in which both the interstitial spaces and interstitial hubs can take place.

As spatial entities, interstitial spaces are composed of functions, qualities, boundaries, and surroundings, and are provided of (or crossed by) different infrastructures that determine their relationality. As well as the built environment, the emergence of interstitial spaces found in sprawling cities are 'followed by processes of delimitation that require some degree of knowledge of the material physical landscape; and practices of demarcation where ditches are dug, boundary markers placed, fences or walls built' (ibid, p. 366). Different from the homogeneous suburban

DOI: 10.4324/9780429320019-8

geography of cities, interstitial spaces are non-standard, diverse, multi-faceted, and heterogeneous in terms of forms and functions. This heterogeneity qualifies the hinterland of cities and influences the way of how interstitial spaces are perceived and interpreted by different actors and institutional representations. Because of their interstitial nature in terms of origins, locations, and functions, the interstitial spaces incite the interests of multiple agencies that make them challenging in terms of governance and its intrinsic politics. Alluding to Koolhaas' architectural spaces where 'everything is done', in the interstitial spaces 'everything is possible' (1995, p. 199). The absolute infilling or the transformation of the interstitial space into a 'place' – the resulting instance of the building-up exercise – is the ultimate change in which the interstitial condition disappears. Also – and considering the context-dependent character of the interstitial spaces – if surroundings disappear, so do the interstitial condition too. Concurrently, if surroundings change, eventually the interstitial condition can (incidentally) re-emerge. This transition defines the temporalities of interstitial spaces and their condition as 'pending' (and 'de-pending') entities while being always 'latent' or while transitioning to become something else. In this transition, the spatial characteristics – and their associated morphologies – change in accordance to the changes of surroundings, surroundings that can be immediate places, transitional areas typically found in sprawling cities, or remote lands located far afield. There is no interstitiality without surroundings or boundaries.

Accepting that cities and regions are not only composed of built-up areas but also interstitial spaces, clarifying what the implications of interstitial spaces are for the extant theories about the city becomes important. Also, accepting that it is not enough to indicate that today 'everything is urban' or 'everything is city' – as some of the planetary urbanisation insights or regionalisation theories suggest – it is important to indicate that we live in an era of 'degrees' (of urbanisation and interstitiality) that differ between regions, between eras, and between the interpretative lenses defined by our epistemological premises about cities and the urban. The observation that an interstitial space can be determined by its infrastructural elements (such as the case of the Cerrillos Airport site), or its functional qualities (such as the orchards of La Pintana), or its institutional arrangements (such as the military airbase El Bosque), or the amalgamation of different (and differing) land uses found in the conurbation spaces, forces interdisciplinarity and the need to overcome binomial categories in the studies of urbanisation and cities. Interstitial spaces are not 'opposed' to cities, and indeed, the one cannot exist without the other: their relation is more symbiotic than dialectic. In that sense, it is important to vindicate the historical role of 'cities' while scrutinising 'the urban' and internalising the interstitiality as a key component of both.

This chapter addresses the implications of interstitial spaces in urban studies. This is structured by fundamental questions that help in clarifying why there could be a focus on interstitial spaces as subject of the extant ideas about the urbanisation process and cities and what the practical implications can be for the planning cities. In the first section, the importance of defining the interstitial space is further reinforced. This sheds light on understanding the influence of interstitial spaces in defining the built environment. Then, the relevance of interstitial spaces for the overall fabric of sprawling urbanisation is discussed. This situates the interstitial spaces within debates of planetary urbanisation and 'the urban question' (Brenner, 2000; Brenner & Schmid, 2014) while drawing upon the ontologies around the interstitial space in the lexicon of urban studies. Finally, a reflection around what makes the interstitial spaces 'urban' and the role of cities as the main entities that shape the urban condition is placed. This is not a blind advocacy on cities, but a reflection that acknowledges the fact that interstitial spaces emerge – both as abstract and concrete entities – because of the existence of cities; a reflection that is perhaps only possible to be made today in which the repeated – and almost *cliché* – mantra that we live in an urban era makes more sense than ever.

8.2 Why the interstitial spaces in urban studies?

> The urban fabric grows, extends its borders, corrodes the residue of agrarian life. This expression, "urban fabric", does not narrowly define the built world of cities but all manifestations of the dominance of the city over the country. In this sense, a vacation home, a highway, a supermarket in the countryside are all part of the urban fabric.
>
> (Lefebvre, 2003, p. 4)

The interstitial spaces are part of the sprawling geography of city-regions; regions that draw together suburbs, satellite towns, and cities into polycentric agglomerations such as or the Randstad in the Netherlands (Lambregts et al., 2012), or scattered urban forms like the ones observed in America, Africa, Latin America, China, or Australia (Chhetri et al., 2013; Cobbinah & Aboagye, 2017; Hamel & Keil, 2015; Rojas et al., 2013). The emergence of these increasingly spatially diffuse forms of agglomeration raises questions for urban studies regarding what lies between developed areas – the interstitial spaces – and the interstitial hubs produced alongside the urbanisation process. As Phelps (2004) noted 'these diffuse forms of agglomeration are notable for throwing up rather anonymous "intermediate" locations or places that are nevertheless important in economic terms' (p. 972). Referring to these intermediate hubs as 'banal locations', Phelps (2004) notably addresses the classical themes of agglomeration theory in explaining today's sprawling urban forms and highlights the economic significance of these intermediate hubs; locations that speak about the interstitial spaces as the context where the 'geography of the banal' is articulated (ibid, p. 973). The author added 'that part of the value of analysing the economic basis of largely overlooked "banal" intermediate places lies in what they may reveal about the functioning of diffuse forms of agglomeration' (ibid, p. 971).

From an urban political ecology perspective, Arboleda (2016) also argues for expanding the study of agglomerations to include those enclaves that lie in the regional and remote interstitial spaces considering their strong influence on the morphology and functionality of cities. Based on the study of mining sites in Chile, Arboleda (2020) indicates that the geopolitics of planetary mining influences the performance of remote cities in the USA. The author argues that 'modern cities are technologically, philosophically, and economically the "inverted mines" of distant resource hinterlands: mineral wealth excavated from the bowels of the earth and then fixed in the urban built environment' (p. 31). In a similar vein, it has been extensively demonstrated that the sprawling growth of many manufacturing towns and cities in China result from a consequent impoverishment (or shadowing process) and destruction of livelihoods in the countryside and other interstitial spaces of remoteness from where the raw material is extracted (ibid). The expansion of iron ore, coal, copper, gas, wood, and other extractive hubs takes place in those regional spaces between cities and regions that influence the wealth of nearby and remote cities alike. This is also the case of the new-tech corporations located in remote lands that create a significant proportion of wealth (and knowledge), which is then used by urban citizens (Phelps, 2015). Reflecting upon the UK, Harrison and Heley (2015) confirm that '57% of net aggregate growth in the UK was accounted for by "intermediate regions" in the period of 1995-2007' (p. 1115). The interstitial hubs therefore take part of the mosaic of interconnected city-regions (Frey, 2019) and operate as interconnected nodes that compete and cooperate with one another across the globe. It has been clarified that city-regions contribute to a substantial part of the global GDP (Rickards et al., 2016), although further research is needed to clarify what of this global wealth solely rest on the interstitial hubs or how 'banal' their economic geography is (Phelps, 2004).

If we pay attention to this literature, however, we must accept that there is a strong focus on – again – what has been done or built (the enclaves or hubs) rather than what is left in-between. This is also what Lefebvre identifies as 'a vacation home, a highway, a supermarket in the countryside' ([1970] 2003:4), all interstitial hubs placed in the interstitial space identified here as 'the countryside'. The focus on the interstitial hubs unveils the still city-centred (or place-centred) lens to analyse the urban; a lens that is probably the main advance in (indirectly) entailing what lies beyond cities and the production of the space (Scott, 2019), but also the main constraint in understanding the interstitial spaces as a wider scope where different forms of accumulation emerge only as one compositional feature.

A key gateway to open this further, however, precisely relates to the interstitial hubs and their interconnections; as discussed in Chapter 6, a sight onto the relational space in which exchanges occur while mediated by boundary infrastructures reveals the existence of the interstitial spaces (Amr, 2021). The relationality of the interstitial spaces becomes a key feature to draw attention to the characteristics of interstitial spaces as realms that facilitate exchange, the relationality that defines the nature of the interstitial condition. This relationality is further highlighted by the ecologies of the interstitial spaces as the wider scope of interlinked environmental processes with local and global salience. As such, the interstitial spaces cannot be reduced to mere locations of extractive enclaves. While assuming 'the reality of our contemporary large cities and agglomerations as a continuum of built-up areas and open space, connected by a network of paths of different sizes and character' (Sieverts, 2003, p. viii), the relationality of these open interstitial spaces becomes key in understanding the urban condition of cities and interstitial spaces alike. This urban condition is not reduced to the scope of cities but expanded 'through' the interstitial spaces, finally contributing to understanding what makes cities 'urban'. The relational character of interstitial spaces ties this potential and cities together by the cascading interdependencies in which one influences the other. We must therefore accept that in the absence of cities, only the urban condition can eventually remain within an 'urban land nexus' (Scott & Storper, 2015; Storper & Scott, 2016); an urban land nexus that is imbricated with the inherent societal progress triggered by the exchange properly found in agglomerations; a space where the interstitiality as such disappears.

This distinction between 'the urban' and 'the city' has been a concern for the disciplines committed to the built environment since time ago, and it is clear today that '... concepts such as "the city" or "the urban" appear to have more and more intellectual labor to perform' (Davidson & Iveson, 2015, p. 647). Mayer in 1969 directly enquired, 'what is a city'? and 'what is meant by urban?' while framing a sort of unambiguous definition for the analysis of cities with international significance (Mayer, 1969). Since there, it has been clarified that if the 'urban' is a process rather than a physical observable entity – and if it is characterised specifically by agglomeration and connectivity – then interstitial space is a significant realm in which the urban condition unfolds. The interstitial geography and its spatial composition may be peripheral to and in between cities, but it is central to the urban. As argued by Zhang and Grydehøj (2021), the case of Zhoushan further suggests that certain core urban functions tend to occur at interstitial spaces; 'urban island studies research suggests that near-shore islands are particularly likely to act as interstices' (p. 2169).

What imbues the interstitial spaces with an essentially urban character is their functional recomposition within a polarised spatial lattice. This polarisation relates to the condition of interstitial spaces as the open and flexible scope of economic, social, and environmental inter-relations crossed by multiple imaginations. This clearly contrasts with the fixed agglomerated tissue of locations and normative constructions that constitute the economic and social essence of the city. As Koolhaas (1995) noted 'where there is nothing, everything is possible' (p. 199), alluding to the idea that the built-up space somehow fixes – or restrains – the opportunities to display further components of the urban, or deploy the potential of the 'creative city' (O'Connor et al., 2020); a claim against the fixing terms of architectural definition of the built-up space while the interstitial space is open to new relationalities and possibilities. This is the context that acts as repository of the 'post-human' buildings; the architecture that Koolhaas identifies as automated gigantic artifacts mostly run by machines; '... a brave new architecture to be realized in record time, without bureaucratic interference' (Koolhaas, 2021, p. 272). In his observation, Koolhaas provides a sense of the type of landscape that these buildings of the remote interstitial spaces created by indicating that 'its boredom is hypnotic, its banality is breathtaking' (ibid, p. 272). The verification that there is not 'bureaucratic interference' – while these structures are nevertheless planned – is somehow the confirmation that a yet unknown (or ignored) form of planning and aesthetics can emerge in the interstitial spaces. A similar observation is made by Brenner and Elden (2009) while commenting on the characteristics of Lefebvre's abstract space as 'it provides a framework for interlinking economic, bureaucratic and military forms of strategic intervention' (p. 359) that overpasses the scale of every state to reach the international and worldwide scale of the planetary state system' (ibid). As such, the interstitial spaces are the realm of planning practices that follow alternative trends and dynamics of the urban; a realm in which 'territorial practices would be the physical, material spaces of state territory, from the borders, fences, walls and barriers erected to mark its external limits, to the creation and maintenance of large-scale infrastructure enabling flows of people, goods, energy and information' (ibid, p. 365).

While being a by-product of the capitalist city, the interstitial spaces are subdued to almost the same structural components of city-regions (Scott, 2019), and thus, would become productive spaces defined by interstitial hubs of different sorts where goods and services are generated and exchanged. The interstitial spaces are also spaces of circulation, movement, and communication through the entire fabric of the interstitial geography and cities. The latter speaks about the extent of the relational character of interstitial spaces and their temporalities while transitioning to become what is often signalised as a 'social space' (ibid). In that sense, the interstitial spaces irrigate from the very core of cities in a matted, tangled, and intricated network of marginalised lands. While the city expands, then the interstitial spaces proceed to configure a sprawling geography that dissipates in the extensive margins where housing tends to lead the way to the open countryside. Finally, the interstitial spaces reach the wider geography beyond and between cities and regions where global processes of accumulation also take place. As such, the interstitial spaces are distinguishable – in quantitative and qualitative terms – from the built-up space by their condition as in-between spaces where some of the less visible societal processes occur, but also because they configure a geographical scope where different manifestations of creativity, informality, economic trade, and permanency emerge and cluster together in a more or less dispersed form.

8.3 Filling a hermeneutical injustice

Even when the attempt of placing a definition is in itself a deterministic exercise (Kivunja, 2018), the value of defining the 'interstitial space' is the one that enables an interdisciplinary operationalisation of its analysis (Boon & Van Baalen, 2019) and the further addition of conceptual and empirical elements that complement its understanding and empirical tractability. Regarding the exercise of proposing a definition, it has been clarified that '... any definition has to posit something that isn't what it is you are defining (...). So that in your head any definition is a drawing of a line. We could not live without definitions. We could not communicate' (Featherstone, 2009, p. 417; Massey, 2009). The clarity of these boundary lines is simultaneously the conceptual restrictions. In some way, the definition of the 'interstitial space' can eventually help in consolidating one of the varied concepts used in the literature about cities and processes of urbanisation, and also further theoretical elaborations around the notions of 'the city' and 'the urban'. If we assume that 'a theory is a set of inter-related constructs (concepts), definitions, and propositions that present a systematic view of phenomena by specifying relations among variables, with the purpose of explaining and predicting the phenomena' (Kivunja, 2018, p. 45), so, the 'interstitial space' is one of these constructs that inter-relate with others.

Another point of argument relates to the lack of attention on interstitial spaces in the extant literature on urbanisation and cities. This is clearly a restriction while analysing cities and their sprawling evolution to become city-regions of global salience. This restriction can nevertheless define the ethos that mobilises knowledge about cities and regions. More specifically – and taking the seminal work of Marina Fricker (2007) around 'epistemic injustices' – the conceptual contribution of interstitial spaces can help in reducing the 'hermeneutical injustice' resulting from the lack of conceptual status of interstitial spaces: we do not know (or know little, or in a fragmentary way) about their existence – and so, we cannot recognise them while analysing the built environment – because the concept does not exist or its empirical tractability is yet ambiguous. In that sense the interstitial spaces are not only a geographical scope – or a collection of geographies – but 'an ethos' that emerges as an alternative point of entry in the studies of sprawling urbanisation, other than the built-up space and its derivation into the city-centred approach. As such, the interstitial spaces allow learning about cities 'through' the heterogeneous spectrum of non-urban geographies located at the margins of the orthodoxies in urban studies; a multiplicity that is explicit in contexts of urban sprawl specifically but that does not reside only there. The interstitial spaces reveal 'an ignored realm' (Koolhaas, 2021, p. 2) of knowledge and multiple realities that overpass binary conceptions proposed by positivist planning (Allmendinger, 2002; Sheppard, 2014).

The values of defining the 'interstitial space' rely more on what the definition opens rather than what it closes. If the interstitial space is not defined, why do we have a substantive effort in developing other critical notions such as cities, towns, rural, suburbia, urban sprawl, and periurban as distinctive categories that – although under debate – help in understanding cities and the urbanisation process? These efforts are not mere semantic *entelechy*; a quick exercise of discourse analysis to some prominent statements made by influential researchers reveals the inevitability of assuming some key terms as unequivocal synonyms while mobilising the progression of certain ideas. Rem Koolhaas, for instance, in an attempt to direct the focus of study to the countryside, indicates that 'today, even a new *city* is familiar: a predictable accumulation of roads, towers, icons ... but as soon as we leave *the urban condition* behind us we confront newness and the profoundly unfamiliar' (Koolhaas, 2021, p. 2), treating the notion of 'city' and 'the urban condition' as one and the same thing. Similarly, the imprecisions derived from the indistinctive use of 'interstitial space' and 'interstices' have also led to confusion while referring to built-up entities – such as towns or other clearly densified places – and spaces such as the geographical space between cities. As clarified in this book, while the 'interstitial spaces' is a 'space', an 'interstice' can be a more defined place: a node, or a physical element, or a very sophisticated agglomeration that articulates relations within the interstitial space. These are the cases of 'interstitial islands' (Zhang & Grydehøj, 2021), 'interstitial cities' (Sayın et al., 2020), intermediate (interstitial) enclaves (Phelps, 2004), or 'interstitial landscapes' (Gandy, 2011).

Despite the absence of any sharply defined break, the attempt that has been nevertheless placed in this book makes the interstitial spaces distinguishable as a broad category of urban phenomena by reason of their spatial condition, morphological heterogeneity, scale, relationality, multi-level governance, innovative capacities, modes of use and occupation, and interconnectivity at local, regional and global levels. These features mark them out as calling for scholarly examination – as Sieverts (2003) avers – to the necessary eviction of other kinds of spaces from the built-up realm that can be considered proper 'places'.

8.4 Vindicating cities from the interstitial spaces

Yet urbanisation can hardly be considered without reference to the interstitial spaces, and as such, it has been clarified that interstitial spaces are produced alongside the urbanisation process by following analytical perspectives and normative agendas framed for the built environment. It has also been clarified that the sprawling character of city-regions contribute to the necessary fragmentation that produces interstitial spaces of proximity and transition, while the regional character of cities opens the agenda of studies at regional scales and the scale of remoteness. Thereby, the existence of cities and their contemporary configurations frame the proliferation of interstitial spaces, and thus, it could not be possible to conceive the interstitiality without reference to the *built-up* as constitutive of what we use to understand by 'cities'.

Despite this, the desire to set an urban agenda based on a wide-ranging critique of 'city-centrism' has gained traction in urban studies. While criticising the fact that we live in an 'urban age' that treat 'the city' as a fixed, bounded, and replicable spatial unit, the notion of urbanisation has emerged as an all-embracing concept in which cities are specific manifestations – somehow subjugated to – of the urban condition. As such, urbanisation can eventually produce other 'urban entities' that can even be more gravitational in their contribution to 'the urban age'. The observation that a form of planetary urbanisation has finally taken place by progressing on Lefebvre's thesis of 'the urban revolution' (1970/2003) seeks to depict the 'urban fabric' as a multifaceted realm that stretches beyond 'the city' to embrace the entire planet through constellations of concentrated and sprawling forms of urbanisation (Brenner and Schmid, 2014). This book is aligned with this thesis, except with the unproportionate disdain placed in some intellectual circles that treat cities and their contribution to the urban condition as almost residual; a trend to treat 'the city' as an almost improper and over-privileged object of study to clarify the urban condition. Although the city-oriented focus in urban studies has been questioned in this book, it does not mean that cities are now irrelevant, and thus, it is important to sustain that 'the city' is still a key category for critical urban theory (Davidson & Iveson, 2015).

While highlighting the interstitial spaces as key entities in contributing to the urban condition of sprawling urbanisation processes, it would be reductive to indicate that the city and the interstitial spaces configure a dialectic relation. It has been extensively clarified that in different degrees the one influences the other; processes of urbanisation derive from the creation of cities and other forms of settlement-type organisations that feed the urban condition while producing interstitial spaces that shape the urban from a different angle. It could not be possible, for instance, to affirm that the urban condition totally disappears in a ghost town or any settlement-type entity of reduced functionality. Conversely, it would not be precise to indicate that 'the city' is totally absent from the meeting point between aboriginal communities in the Amazon, while they reach the meeting point through a socially defined route of access and gather around an agreed (and so, planned) shared space of interchange. So, it is clear that Storper and Scott's 'urban land nexus' (2016) is not the city as we know, but taking the city out of the theorisation of the urban is certainly a reductive exercise upon an appreciation of cities as unique contemporary manifestations. For sure the widely repeated mantra that 'now *all* is city' (Koolhaas, 2013, p. 29) can be an exaggeration (although provocative), but looking at the historical exist- ence of well-bounded cities – which have been the scenario of civilisation for centuries – with a certain intellectual disdain is an exaggeration too. This also entails a rejection of 'a non-urban "outside" to cities, (...) in favor of a concept of "planetary urbanization"' (Davidson & Iveson, 2015, p. 651), whose urban fabric even include those desertic geographies where there is no presence of any societal process at all. This is why the Sahara Desert is more an interstitial 'space' of remoteness rather than an in-between 'place'; the Sahara is not possible to be assimilated as 'equal' to those regional spaces, hubs, and settlements fully immersed in the transformation of north-east African societies historically dependent of the Mediterranean Sea and the Nile river. Lefebvre (1970/2003) has indeed clarified this, argu- ing that although 'the *urban fabric* grows, extend its borders, corrodes the residue of agrarian fife (...) the only regions untouched by it are those that are stagnant or dying, those that are given to "nature"' (Lefebvre, 2003, pp. 3–4). This is why Davidson and Iveson (2015) question: 'If the extended geographies of urbanization have no necessary or privileged or significant relation to cities, then what makes them urban?' (p. 655). These interstitial geographies and the sprawling urbanisation of global gravitation can only be meaningful to be urban in relation to cities.

The fact that interstitial spaces influence the urbanisation process is more a direct role of the interstitial hubs, and by extension the interstitial spaces that appear as the context where the hubs emerge. This implies that some interstitial spaces – in the absence of interstitial hubs or with very diminished relationality (like the Sahara Desert) – can definitively be less influential and indeed more 'influence by' surrounding cities that instrumentalise their heterotopic character. This confirms the context- dependent nature of interstitial spaces while placing the need of extending urban theory over the interstitial spaces to investigate the extent to which they describe a distinctive social system around cities or if cities as social systems spill into the interstitial geography. This biunivocal relation is possible to be observed in the interstitial spaces at all scales and levels of relationality. Definitively, while it is possible to agree with the notion that the urban can extend beyond the realm of 'the city' and irrigate into the interstitial spaces, the assumption that 'the city' disappears as a specific form of urbanisation – or that cities do not mobilise urban thinking –

would be conceptually and empirically erroneous. The interstitial spaces do not emerge by themselves but precisely 'because of' the presence and the sprawling nature of cities. The work of scrutinising 'the urban' solely based on the interstitial spaces and without any reference to 'cities' would be found difficult if not impossible. It does not mean, however, that 'the urban' is the same as 'the city', and/or that 'the urban' is solely contained within cities.

8.5 Conclusions

As well as there are different types of urban agglomerations – that can be recognised in different degrees as 'cities', of 'less city', or 'less urban' – interstitial spaces are not absolute entities characterised by being completely empty, unfunctional, or inert. Like any urban agglomeration, there are also levels of interstitiality in which 'the degree of emptiness' differs from one interstitial space to another, and from region to region. Measuring the (degree of) emptiness of an interstitial space can be a matter of further research, as well as measuring the extent of urbanisation of cities has been the focus in urban studies and planning since the years after the industrial revolution.

The implications of interstitial spaces have salience at theoretical and empirical level\s, while the concept itself serves in clarifying 'what is' and 'what is not' an interstitial space and how they make cities urban. Further theoretical discussion, however, is needed to understand the way of how interstitial spaces contribute to the urban condition beyond their spatial, environmental, and geographical characteristics. The theoretical gateways from anthropological studies can eventually support reflections in this way and help to disclose the potentially ambivalent condition of interstitial spaces as 'places' while intensifying their functional responses to the urban.

In practical terms, it has been clarified that interstitial spaces offer a new realm for planning practice, spaces in which planning orthodoxies are somehow absent or have simply been shifted to satisfy the opportunities framed by the interstitiality at local, regional, and global scales. These opportunities relate to the emergence of new types of interstitial hubs – or Koolhaas' post-human architecture – that suggests new approaches to aesthetics, functionality, material exchange, and social interaction. Just like the accompanying lexicon of our understanding of cities, all these categories are not probably needed to be redefined but rather expanded – if hermeneutically possible – to include and embrace those manifestations of the urban occurred in the interstitial spaces. If not, there is a chance to definitively embark on the adventure of redefining the scope, imaginaries, and ontologies of planning and urban studies (or the 'studies of the urban') and assuming the conceptually and empirically disruptive character of the interstitial spaces.

References

Allmendinger, P. (2002). Towards a post-positivist typology of planning theory. *Planning Theory*, *1*(1), 77–99.

Amr, A. A. (2021). Underlying relational dimensions of flow transitions along ring roads and their impacts on the typo-morphology of open spaces: Two cases from Nordic countries. *Urban, Planning and Transport Research*, 1–29.

Arboleda, M. (2016). In the nature of the non-city: Expanded infrastructural networks and the political ecology of planetary urbanisation. *Antipode*, *48*(2), 233–251.

Arboleda, M. (2020). *Planetary mine. Territories of extraction under late capitalism*. Verso.

Arvanitis, E., Yelland, N. J., & Kiprianos, P. (2019). Liminal spaces of temporary dwellings: Transitioning to new lives in times of crisis. *Journal of Research in Childhood Education*, *33*(1), 134–144.

Boon, M., & Van Baalen, S. (2019). Epistemology for interdisciplinary research–shifting philosophical paradigms of science. *European Journal for Philosophy of Science*, *9*(1), 1–28.

Brenner, N. (2000). The urban question: Reflections on Henri Lefebvre, urban theory and the politics of scale. *International Journal of Urban and Regional Research*, *24*(2), 361–378.

Brenner, N., & Elden, S. (2009). Henri Lefebvre on state, space, territory. *International Political Sociology*, *3*(4), 353–377.

Brenner, N., & Schmid, C. (2014). The 'urban age' in question. *International Journal of Urban and Regional Research*, *38*(3), 731–755.

Chhetri, P., Han, J. H., Chandra, S., & Corcoran, J. (2013). Mapping urban residential density patterns: Compact city model in Melbourne Australia. *City, Culture and Society*, *4*(2), 77–85.

Cobbinah, P. B., & Aboagye, H. N. (2017). A Ghanaian twist to urban sprawl. *Land Use Policy*, *61*, 231–241.

Davidson, M., & Iveson, K. (2015). Beyond city limits: A conceptual and political defense of 'the city' as an anchoring concept for critical urban theory. *City*, *19*(5), 646–664.

Featherstone, D. (2009). The possibilities of a politics of place beyond place? A conversation with Doreen Massey. *Scottish Geographical Journal*, *125*(3–4), 401–420.

Frey, K. (2019). Global city-region. In A. Orum (Ed.), The Wiley Blackwell Encyclopedia of Urban and Regional Studies. 1–6. John Wiley & Sons, Ltd.

Fricker, M. (2007). *Epistemic injustice. Power & ethics of knowing*. Oxford University Press.

Gandy, M. (2011). Interstitial landscapes: Reflections on a Berlin corner. In M. Gandy (Ed.), *Urban constellations* (pp. 149–152). Jovis.

Hamel, P., & Keil, R. (Eds.). (2015). *Suburban governance: A global view*. University of Toronto Press.

Harrison, J., & Heley, J. (2015). Governing beyond the metropolis: Placing the rural in city-region development. *Urban Studies*, *52*(6), 1113–1133.

Kivunja, C. (2018). Distinguishing between theory, theoretical framework, and conceptual framework: A systematic review of lessons from the field. *International Journal of Higher Education*, *7*(6), 44–53.

Koolhaas, R. (1995). Imagining the nothingness. In R. Koolhaas, & B. Mau (Eds.), *S, x, l, XL*. The Monacelli Press.

Koolhaas, R. (2013). 'Atlanta'. In R. El-Khoury, & E. Robinns (Eds.), *Shaping the city. Studies in history theory and urban design* (2nd ed., pp. 23–31). Routledge.

Koolhaas, R. (2021). *Countryside. A report. The countryside in your pocket!* Taschen.

Lambregts, B., Kloosterman, R., van der Werff, M., Röling, R., & Kapoen, L. (2012). Randstad Holland: Multiple faces of a polycentric role model. In Sir P. Hall & K. Pain (Eds.), *The polycentric metropolis* (pp. 147–155). Routledge.

Lefebvre, H. (2003). *The urban revolution.* University of Minnesota Press.

Mayer, H. M. (1969). Cities and urban geography. *Journal of Geography, 68*(1), 6–19.

O'Connor, J., Gu, X., & Lim, M. K. (2020). Creative cities, creative classes and the global modern. *City, Culture and Society, 21*, 1–6.

Phelps, N. A. (2004). Clusters, dispersion and the spaces in between: For an economic geography of the banal. *Urban Studies, 41*(5–6), 971–989.

Phelps, N. A. (2015). *Sequel to suburbia: Glimpses of America's post-suburban future.* MIT Press.

Rickards, L., Gleeson, B., Boyle, M., & O'Callaghan, C. (2016). Urban studies after the age of the city. *Urban Studies, 53*(8), 1523–1541.

Rojas, C., Mu, I., & Pino, J. (2013). Understanding the urban sprawl in the mid-size Latin American cities through the urban form: Analysis of the Concepción metropolitan area (Chile). *Journal of Geographic Information Systems, 5*, 222–234.

Sayın, Ö, Hoyler, M., & Harrison, J. (2020). Doing comparative urbanism differently: Conjunctural cities and the stress-testing of urban theory. *Urban Studies,* 1–8.

Scott, A. J. (2019). City-regions reconsidered. *Environment and Planning A: Economy and space, 51*(3), 554–580.

Scott, A. J., & Storper, M. (2015). The nature of cities: The scope and limits of urban theory. *International Journal of Urban and Regional Research, 39*(1), 1–15.

Sheppard, E. (2014). We have never been positivist. *Urban Geography, 35*(5), 636–644.

Sieverts, T. (2003). *Cities without cities. An interpretation of the Zwischenstadt.* Spon Press.

Storper, M., & Scott, A. J. (2016). Current debates in urban theory: A critical assessment. *Urban Studies, 53*(6), 1114–1136.

Zhang, H., & Grydehøj, A. (2021). Locating the interstitial island: Integration of Zhoushan Archipelago into the Yangtze River Delta urban agglomeration. *Urban Studies, 58*(10), 2157–2173.

9   Conclusions

This book opens with a city that was, symbolically, a world: it closes with a world that has become, in many practical aspects, a city.

(Mumford, 1961, p. xi)

9.1 Introduction

In the preceding chapters, the nature of interstitial spaces as part of the sprawling urbanisation of cities and regions has been clarified. The origins of the interstitiality at the core of Sieverts' *Zwischenstadt* has been illustrative to understand the interstitial spaces that are produced alongside the urbanisation processes; a process that has reached the status of sprawling city-regions, '...a completely different and new form of the city, which is spreading across the world: the urbanised landscape or the landscaped city' (Sieverts, 2003, p. xi). Although the involved notions and their empirical observables have been further developed (Bruegmann, 2005; Hamel & Keil, 2015; Phelps & Wood, 2011; Scott, 2019), Sieverts' work has been seminal for this book to situate the geographical scope of urban sprawl and its multifaceted character as equally composed of built-up lands and interstitial spaces. This simple (and fundamental) locational insight has been relevant – not only in geographical terms – to understand the implications of interstitial spaces in contemporary debates around the urbanisation process, a process of global salience that has ignited critical enquiries around the role of cities and their urban character.

Somehow, '...the widely repeated mantra that more than half the world's population now lives in cities has given rise to a plethora of research seeking to understand the urbanized human condition' (Davidson & Iveson, 2015, p. 648) – while the numbers really indicate that 'if the majority of the world's population is now in officially defined urban areas, then the majority of that population is actually suburban' (Phelps, 2021b, p. 353) – confirms the importance of revisiting Sieverts' *Zwischenstadt* through the lenses of city-regions and their sprawling character as critical elements of an era of planetary (sub)urbanisation. The *Zwischenstadt* suggests 'a

DOI: 10.4324/9780429320019-9

reflection of the widespread emergence of large spatially extended urbanized areas all over the world, each of them locationally anchored by one or more metropolitan centres, and each of them spreading far outwards into diffuse hinterlands comprising mixes of agricultural land, suburban tract housing, miscellaneous industrial and commercial properties, local service centres, and subordinate urban settlements' (Scott, 2019, pp. 554–555). This 'diffuse hinterland' is precisely the geography of the interstitial spaces, a geography where 'the nodes and settlements are connected by a continuous network of transportation routes and paths of different types and capacities' (Sieverts, 2003, p. ix). These relational aspects entail the nature of interstitial spaces as relational spaces, a relationality that covers from the liminal instances of proximity found in the inner suburbia, up to the very large spaces of global exchange. These scales were also elicited by Sieverts' work while defining the *Zwischenstadt* as 'the type of built-up area that is between the old historical city centres and the non-places of movement, between small local economic cycles and the dependency on the world market (2003: xi). In this regard, enough work has been done in describing the magnitudes of city-regions and how their sprawling growth has overpassed any attempt of land management at local levels. City-regions have been treated as the 'urban behemoths' (Scott, 2019, p. 555), with special reference to the logic of agglomeration, growth, and spatial interaction. However, these 'behemoths' are somehow still timid entities when compared with the colossal magnitudes of the interstitial spaces of remoteness or the impacts of some interstitial hubs on climate change and the skyline of remote cities far afield. Without a doubt, indeed, it is possible to assert that the interstitial spaces of remoteness contain these 'behemoths' as part of its colossal fabric, while the same interstitial fabric penetrates the city up to its very core. In some cases, these penetrations are morphologically clearer and continuous – such as the green wedges observed in Copenhagen or the Lagan River in Belfast – and in some others are more intricate, fragmented, and randomly distributed – such as the interstitial spaces observed in cities like Santiago de Chile.

As proposed, the salience of defining the interstitial spaces relies on the importance of identifying and handling a sort of *slippery object of study* while researching cities and what makes them 'urban' in their transition to consolidate their sprawling character. The composed term 'interstitial space' – as well as 'city-region' or 'non-places' to mention a few – does not fall within a culturally assimilated and automatic meaning and requires conceptual scrutiny and ontological dissection to explore its theoretical and empirical tractability. This exercise would not have been possible without the seminal work of the intellectual giants – Lefebvre, Augé, and Foucault – who previously elaborated around the notions of 'space' and 'place'. Their interdisciplinary insights around these fundamental terms gave space to the 'interstitial space' to be part of the studies of cities and the urbanisation process. These conceptual grounds along with the nature, the scales, and relationalities of interstitial spaces allow the operationalisation of their analysis and suggest an agenda of research for a dimension of the built environment that has been treated in a very fragmentary way. The 'interstitial space' can serve as one of the axioms used to elaborate around the usual readings of the urban in most theories of the city while simplifying and situating the geography of the 'in-between' in a less hermetic and generic way. As seen through the cases of Santiago de Chile, the operationalisation of the interstitial spaces allows the analysis of concrete problems around issues of extended suburbanisation, infrastructure, land fragmentation, urban regeneration, governance, and others. This is important in theoretical terms too considering that 'theory is not to regress infinitely into meta-theoretical obscurity with laments that "the world is more complex than that"; dialectical method is to be fully utilized; theoretical concepts are to be verified empirically; and theory is to have relevance to pressing practical urban planning problems' (Phelps, 2021b, p. 353). In this book, a perspective of this entire agenda of investigation and an elaborated reflection on some of the pressing debates on cities and urbanisation have arisen in regard to the origins, nature, and theoretical status of interstitial spaces. As discussed, the extant literature on interstitial spaces uses a variety of terms, is fragmented, and is quite singular in its treatment of interstitial spaces in architectural, ecological, or other terms. As demonstrated in this book, however, the notion of 'interstitial space' appears as more embracing as it encapsulates the multiple physical, geographical, spatial, functional, economic, and temporary dimensions of the geographies of the 'in-between'.

The multifaceted nature of the interstitial spaces speaks about the potentials of land fragmentation and the spatial juxtapositions, hybridisation, intersections, embeddedness, and multiple articulations defined by their intermingled infrastructures of relationality. These are creative scenarios for innovations in planning, in which the exercise of extending what has been done for producing 'the city' – or keeping the 'business as usual' – can potentially be another 'zombie approach'. Long-cherished planning policies such as construction norms, zoning, aesthetical views, historical heritage, or greenbelts inter alia can fail while being implementing in a realm that – without reflection or reconsideration – questions the very ethos of planning as a prescriptive discipline. The potential of interstitial spaces is separated from the positivist thinking that has dominated the planning of cities and calls for the development of innovation and creativity on land fragmentation beyond bi-dimensional conceptions of land use. In that sense, the efficacy of interstitial spaces in being produced at the margins of cities and the urbanisation process suggests that rejected ontologies of 'the interstice' – such as 'impurity' or 'strangeness' while being elements 'out of the norm' – can activate alternative views to the supposed ugliness, emptiness, abandonees, and marginality of interstitial spaces. What are a mining site and its roads, circulations, pollution, excavations, tunnels, infrastructures, and other elements that support its extractives functions in the outskirts of a city if not an ugly space? What is a conurbation, if not a space out of norms and plans?

9.2 The lessons from Santiago de Chile

Santiago de Chile is a city-region that has been the subject of fragmented urban expansion linked to both 'autoconstruction' and privatised construction of housing and infrastructure in the past four decades with important consequences for social segregation. In this book, this case is used as a basis for abstraction and generalisation, though the agenda of research on interstitial spaces is one that has salience to the vast majority of cities. Santiago de Chile, however, provides a good example of the significance of interstitial spaces in the context of urban sprawl specifically. Through the case of Santiago, it is also clear that although planning policies tend to be increasingly standardised, Santiago's interstitial spaces are diverse and non-homogeneous elements. They also carry different connotations among actors depending on their social, environmental, and distinctive levels of functional integration. While social-based actors see these interstitial spaces as suitable for public facilities and services, policy-makers identify them as suitable for social housing. Similarly, developers perceive Santiago's interstitial spaces as underused lands that should be developed or further densified. Designers see the opportunities for innovative practices in architecture. All in all, the evidence suggests that apart from their origins, role, and future land uses, interstitial spaces are produced at the margins of planning policies for the city and, thus, determined by specific social, economic, and political circumstances. All these elements confirm the context-dependent nature of the interstitial spaces, a dependency that influences their strategic position in the sprawling expansion of Santiago, their meanings, significance, and political interpretation as opportunities for radical urban transformations.

The regional interstitial spaces of Santiago rarely relate to single territorial units. They encompass wider territories of multiple levels of administration and interests. These are the cases of the Cerrillos airport site or the gravel pits in the boundary area of La Florida and Puente Alto or the southern conurbations of Maipu/Padre Hurtado and San Bernardo/Lo Herrera. These regional interstitial spaces suppose inter-municipal and local-central coordination that supposes a major challenge for the Chilean planning system. For the case of the gravel pits, the municipality of La Florida has a clear interest in their reconversion, but the municipality of Puente Alto is yet indifferent in front of the privatised property regime of the area. The case of the Cerrillos airport site and the CBD project is under constant political peril, which demonstrates the instability of interstitial spaces while applying normative planning approaches and project-based rationales for their urbanisation. The case of La Platina demonstrates the marginalisation of interstitial spaces even when at least three institutional representations operate upon their current land uses, maintenance, and future destinations: the MINVU, the Ministry of Agriculture, and the Municipality of La Pintana. As a way of solution, the site is again subject of project-based consensus and integrated into the wider inter-communal network of potential green areas of the 'South Park' project. A similar situation is observed in the case of the military base El Bosque, in which the site is in stalemate considering disparities between military, central, and local authorities. The conurbations describe unregulated encroachments on rural lands tensioned by attributions of different ministries and urban/rural regulations. In this context, centralised decision-making takes advantage of unregulated conditions of the conurbation space to implement infrastructure projects of regional significance.

The heterogeneity of interstitial spaces in Santiago is also a common characteristic. Despite standardisations in planning, Santiago's interstitial spaces emerge from different institutional purposes – agricultural, infrastructural, others – that influence their heterogeneity at different levels. This heterogeneity differentiates the interstitial spaces in comparison to their homogenous residential surroundings while posing planning and design challenges against standardisation. In this heterogeneity rests the different relationalities that influence the spatial and functional integration of Santiago's interstitial spaces. The orchards of La Pintana, for instance, have adapted their functions – despite radical political changes and pressing policies to urbanised them – to remain integrated at local and regional levels; adaptations that are concurrently occurring along with the transformations of their surroundings. Conversely, the gravel pits of La Florida and Puente Alto describe a lower integration due to their functions, infrastructural, and spatial aspects.

Based on this evidence, the most critical aspect regarding the role and potential of interstitial spaces in Santiago de Chile is political. The common assumption that Santiago's interstitial spaces are marginalised entities – and thus 'out of the norms' – mobilises planning thinking around their urbanisation or other forms of reintegration into the city's fabric. In particular, their land capacity, location, and connectivity infrastructure place them as strategic and suitable for triggering radical transformations at local and regional levels. More importantly, the analysed interstitial spaces appear as politically sensitive considering their residential and communal surroundings characterised by the concentration of low socio-economic groups. Their political relevance also relates to the fact that the southern space has been consolidated as the main axis for the sprawling expansion of the city; a consolidation marked by the concentration of poverty over the last four decades in order to sustain the application of social benefits from an insipid – and somehow agonic – Chilean welfare state.

Santiago's interstitial spaces emerge as scenarios of differing interests tensioned by narratives of preservation and change. While developers understand the interstitial spaces as underused, undeveloped (or less-developed) and necessary to be functionally intensified, central authorities and policy-makers see Santiago's interstitial spaces as spatial barriers – non-relational – that interrupt the continuous urban fabric and its sense of unity; spaces that 'must' be urbanised and transformed into well-located social housing developments. Local residents contest these views stepping on narratives of 'the right to the city' and 'environmental justice' while signalising interstitial spaces as the necessary open tracts for leisure and social encounters that reinforce their sense of community. Finally, local planners use the interstitial spaces to mobilise their hopes of improving the communal urban quality and change the socio-economic profile of their communal hinterlands while defending the interstitial spaces as opportunities to provide services and green space. This lack of consensus influences not only their interpretation (what the interstitial spaces are) but also their futures (what the interstitial spaces could be). In this vein, if there is a single commonality around the interstitial spaces of Santiago is their suitability for the implementation of urban 'projects'. This highlights a substantial difference in the way of how institutional actors understand the transition of interstitial spaces: there are 'policies for the city' but 'projects for the interstitial spaces'. This surely draws the attention of multiple and politically mediated imaginaries – above all from an urban design perspective – and elaborates on the ideologies upon the type of sprawling suburbia that the interstitial spaces suggest.

Santiago's interstitial spaces are non-standard and active elements of the urban fabric and are perceived in different ways by different actors. As such, the interstitial spaces are contested realms that, although marginalised, trigger the illusion of being reclaimed for further urbanisation. They are also spaces of resistance and preservation of social, environmental, political, and economic values and make clear the fact that – regardless of their differential visibility in the sprawling growth of Santiago – normative approaches in planning do not have the tools to codify the interstitial spaces as part of the development agenda. What the interstitial spaces of Santiago's urban sprawl finally suggest is that politics – as seen in the seven cases – is never absent from the fortunes of interstitial spaces as much as the built environment. As such, their conceptual grounds, nature, scale, and relationality have implications for the extant theories of the urban and cities.

9.3 Final remarks

The evolution of city-regions has defined this encounter between the production of the built-up space and the interstitial spaces, an encounter that illustrates the necessary contrast to understand the interstitiality as a distinctive category in the processes of sprawling suburbanisation with global significance. This encounter occurs "thanks to cities' and not despite them, and thus, the assumption that cities must be increasingly dismissed from the extant theories about the urban is not only erroneous but also theoretically and empirically misleading. It has been demonstrated in this book that processes of urbanisation and the sprawling development of city-regions trigger the emerge of interstitial spaces. As such, their origins lie at the core of planning rationales for the production of the space; a process that has reached the status of being global – or planetary – and in which 'cities' – and other settlement-type agglomerations – are another component of the wide global urban fabric as much as the interstitial spaces. In this fabric, the interstitial spaces describe not only an ignored geographical scope but also an ignored realm of knowledge, a realm that cannot be ignored from the extant reading of the urban while theorising about cities. This 'ignorance' is of course contested ('who' ignores it!) as planning the city does not imply that the interstitial spaces are not planned in one way or another. Phelps (2021a) has noted that planning must not be confused with the institutionally framed set of formal plans, the activity of (what we identify as) professional planners, or normative agendas placed by our validated institutions, as 'urban planning and its associated imagination (...) are prescriptive about who is doing planning' (2). It may even be more apposite to ask the question 'whose city?' of interstitial spaces given the competing interests surrounding them and the fact that in some way or another they are not totally 'inert'. In that sense, if in the built-up space we analyse the characteristics of the 'built environment', it follows that in the interstitial spaces, we would need to analyse the degree of 'emptiness' – or the 'nothingness' proposed by Koolhaas – and how it differs between interstitial spaces, scales, and regions. This 'nothingness' is probably the most elegant (or ambiguous) way of describing our hermeneutical absences about cities, considering that – again – what lies between and beyond the built-up space is a realm to be further explored; a realm in which 'planning has a geohistory and imagination that far precede planning as a modern profession (Phelps, 2021a, p. 4).

The interstitial spaces are not spaces of traditional consensus derived from symmetric or regulated negotiation. These are contested and politically mediated fields where – as observed in the case of Santiago de Chile – someone 'wins', and someone 'loses'. That partly explains the origins of interstitial spaces, a production of interstitiality that runs in parallel to the production of the built-up space subject to constraints resulting from shared social or regulatory constructions that can be unmade, improved, or reclaimed to allow development. This location of interstitial spaces at the margins of the urbanisation process is also their efficacy: they efficiently emerge out of the sight of our knowledge and our shared cultural constructions about cities. In that sense, their present condition is subject of endless disparities, disparities that nevertheless illustrate the inefficacy of the urbanisation ethos around 'filling all the gaps' as the interstitial spaces go nevertheless rampant from the very core of cities up to the global spaces of remoteness.

Despite standard policies for promoting urban growth and determining the production of the space, interstitial spaces are varied and specific and so subversive elements that resist any form of standardisation. It questions a possible extension of our current analytical apparatus on the interstitial spaces, as they challenge our ideas around governance, planning, design, aesthetics, our understandings about their random origins as outcomes of less controlled processes in planning, and the very mechanisms for the production of the space. As such, the interstitial spaces are subject of inspection in their own rights.

References

Bruegmann, R. (2005). *Sprawl: A compact history*. University of Chicago Press.

Davidson, M., & Iveson, K. (2015). Beyond city limits: A conceptual and political defense of 'the city' as an anchoring concept for critical urban theory. *City*, *19*(5), 646–664.

Hamel, P., & Keil, R. (Eds.). (2015). *Suburban governance: A global view*. University of Toronto Press.

Mumford, L. (1961). *The city in history*. Houghton Mifflim Harcourt Publishing Company.

Phelps, N. (2021a). *The urban planning imagination*. Polity Press.

Phelps, N. A. (2021b). Which city? Grounding contemporary urban theory. *Journal of Planning Literature*, 36(3) 345–357.

Phelps, N., & Wood, A. (2011). The new post-suburban politics? *Urban Studies*, *48*(12), 2591–2610.

Scott, A. J. (2019). City-regions reconsidered. *Environment and Planning A: Economy and Space, 51*(3), 554–580.

Sieverts, T. (2003). *Cities without cities. An interpretation of the Zwischenstadt*. Spon Press.

Index

Note: Page number in *italics* represent figures.

Printed and bound by CPI Group (UK) Ltd, Croydon, CR0 4YY

17/10/2024

01775712-0001